Science Magic

Fun Guaranteed!

Sue McGrath
BSc Hons. PGCE CED DASE MEd FInstP

AuthorHouse™
1663 Liberty Drive, Suite 200
Bloomington, IN 47403
www.authorhouse.com
Phone: 1-800-839-8640

AuthorHouse™ UK Ltd.
500 Avebury Boulevard
Central Milton Keynes, MK9 2BE
www.authorhouse.co.uk
Phone: 08001974150

First published by AuthorHouse 8/7/2007

ISBN: 978-1-4259-7061-1 (sc)

Printed in the United States of America
Bloomington, Indiana

This book is printed on acid-free paper.

Bloomington, IN   authorHOUSE®   Milton Keynes, UK

Why Science Magic?

Magic practice is said to be a secret wisdom. As a science, it is the gaining of knowledge. As an art, it is the understanding of the secrets of nature. As a pursuit among human endeavours, its primary focus is the manipulation of physical processes by wielding unseen forces that provide superhuman or supernatural powers. Magicians lead us to believe they have gained their powers due to an advanced learning and profound understanding of nature. Their actions are beyond the capabilities of mere mortals, or muggles, such as us!

The performance of Magic is arguably the most ancient of all human arts. The word *Magic* derives from the Persian *magus* meaning a priest of the Zoroastrian religion. As *magos* in Greek and magus in Latin, it quickly took on the meaning of charlatan in the ancient Pagan world.

Many definitions assume that Magic involves supernatural powers or forces. In the ancient Near East, as in many other cultures, even today, there exists no such expression as natural and supernatural. We live in the world of nature which is the creation of the deity and everything that exists involves the deities to a greater or lesser degree. A force may be outside the human realm, but it is not outside the realm of nature.

"Whether the gods are old or new, whether they come from Egyptian, Mesopotamian, Levantine, Greek, Jewish, or Christian traditions, religion is an awareness of and reaction against human dependency on the unfathomable scramble of energies coming out of the universe...How could ordinary men and women . . . get something out of their lives? It is at this point that Magic became a necessity to the lives of ordinary people. . . from time immemorial Magic has survived throughout history, through the coming and going of entire religions, the scientific and technological revolutions, and the triumphs of modern medicine." Hans Dieter Betz, **The Greek Magical Papyri in Translation**, pp. xlvii-xlviii.

We should remember that science, Magic, and religion form a continuum with no clear boundaries between them. It is a modern trait for us to separate interlinking systems into various systems which in reality are interconnected. You will notice this done on numerous occasions throughout the book whilst dealing with the science surrounding themes such as colour, energy and forces. Magicians are no different, for them to develop within their field they split the teaching of Magic into 6 basic branches:

Manipulation: Tricks with coins, cards and other small objects which, by slight of hand, are made to disappear, change colour or perform other interesting feats.

Comedy: The amount of humour and comedy put into an act depends on the individual, you will find that as your confidence increases more comedy will naturally be added – sometimes you will find it is the tricks which go wrong are the ones which produce the loudest uproars – so stage some to fail and note the reactions.

Mental: I suppose we can say that anybody that is willing to stand up in front of a load of people they don't know and make an idiot of themselves is mental; but true mentalists are Magicians whose tricks appear to predict future events, read minds, find hidden objects and so on.

Escape: Escapologists specialize in escaping from enclosed spaces. They are usually chained up and handcuffed. (Hmm!) This is an area of Magic that I have not yet perfected but believe me there have been times, whilst on stage in front of a very boisterous group of teenagers, I have wished to have the ability to escape down into a very large hole!!

Illusions: Magic on any scale is an illusion but Magicians who specialize in large tricks such as sawing people in half and making their mother-in-laws float out of the house are known as illusionists, and finally

Close-up: Close up Magic is performed at close range at dinner parties, bars and other social gatherings.

I will ask again "Why Science Magic?" Magicians are secretive, they very rarely let you know that the wonder behind the trick has nothing really to do with their advanced learning, any body could repeat the trick if they have the patience to practice, if they only told us how. The wonder is 'knowing' how things work, how minds work and learning to tell a good story. Scientists usually have the first two skills but fail to become good story tellers and hence fail to capture the imagination of those with whom they wish to share their knowledge. So, in a nutshell, Magicians don't share knowledge but tell good stories and get invited to parties. Scientists usually have a very good story to tell but have not developed the artist skills to be able to do so and therefore are ignored at parties! The aim of this book is join the two skills together so that we get a really good story which can then be retold by all who hear and all those that hear will enjoy it so much will want to go out and tell it themselves!

The science tricks in this book could be categorized as manipulations, illusions and close-ups. Many of them have the opportunity for comedy to be woven into the acts. And as with any form of performance whether by a teacher or Magician the key to success is 'practice'. There is just no way around it. You need to pay attention to what you are doing and make a conscious effort to improve – do this regularly and you will!

Another secret to being a good Magician is realizing that the majority of the act is a performance; acting and the telling of tales. A Magician trains to make every word and motion lead his trusting, innocent audience into a completely inescapable and sometimes ridiculous trap.

You would be quite surprised to find out how many Magicians actually have a physics or chemistry background (advanced knowledge and all that!). We all know kids love Magic and to be honest the majority of adults do too. Teaching science should at times be like a Magic show; audience participation must be allowed and the structure of the lesson or activity should contain humour and be tailor made to fit the age and interest of the audience.

So whether you are using this book for family fun, as an aid for science teaching or a means to engage young people to do science, I encourage you to develop a performance so that you can capture the imagination of your wards and in turn get them to develop their own style which will help turn them into fully fledge apprentice Science Magicians!

By doing so you will give them the ability to express themselves, to imagine complicated subjects, create new things and maybe even tap into the hidden meta-physical universal energy in which we live which at present cannot be detected even by the most advanced scientific instrument!

Our goal as educators should be, whilst allowing them to thoroughly enjoy the learning process, to give our youth the skills and processes by which they can describe all nature in physical and theological terms. Modern physics allows us to investigate the unseen aspects of nature; sci-fi novels are really just a collection of modern mythologies, maybe, our future young scientists will be able to combine the above with theology and come up with an explanation of 'meta-physics' which can be understood by all!! Pigs may fly! Go Babe!

By introducing your young wards to the practical side of science, whilst also allowing them to develop their expressive and imaginative powers, you will give them the skills and tools to be able to fully engage in the wonderful world surrounding them. You never know, they may even grow up to become a science hero, one who is able to successfully combine the four fundamental forces of nature with the universal energy force – a force which may emanate from Einstein's suggested alternate dimensional universe which co-exists with our real observable one.

Let the force be with you!

P.S. Did you buy this book because Magic is in the title? Well so long as you become an active reader, Magic is what you will get... the Magic of unadulterated childish fun! Though I must warn you – to truly get the most out of the ensuing pages you must let your imagination run wild. Be the most animated character, leave all societal boundaries behind you... join the world of the Hans Christian Anderson, Roald Dahl, J K Rowling and Richard Feynman; tellers of stories which fire the imagination and lead you down a path of discovery and wonder.

Yes, the key to a successful trick is the story-telling and should I say it? ... Okay... lying as well! No, not just lying, the key to capturing your audience is talking about huge mind boggling and muggling lies! Though it is important that these lies be told enthusiastically and be thought provoking however they must also be told with a straight face and a steady stare!

Many of the activities in this 'Little Book of Magic' will reinforce and extend all the followers of sciences' current knowledge and understanding and will hopefully leave all believing that science is like a huge slice of chocolate cake – enjoyment only cometh when you eateth! Read, consume and enjoy!

Contents

Chapter 1

"May the Force be with You!"

Whilst uttering "May the Force be with you!", cut through the atmosphere with your 'all singing and dancing' Jedi sword. You may even want to wear your voice changing helmet. What you don't have these toys?? Too old? Rubbish! They are fantastic examples of tranducers; the physicists' fancy word for objects, live and animate, which change energy from one form to another – a must for all households and classrooms.

The driving force behind the Star Wars world of exploding planets and intergalactic wars is also the force that is at the centre of many religions such as Taoism; and the same force explored in this book! 'The Force' which gives a Jedi knight his power is the same as the force that we use and abuse in the world we live in; the force we draw upon from the energy field which surrounds us and penetrates us. The force, which keeps our planet in orbit around the Sun; our closest star, which is by the way on its own life journey and, whose discarded energy feeds us. The same force which binds the galaxies together and is also used by many hands-on healers – a mysterious, unidentified force – a force which poses many questions and is open to you all to study and provide answers...

Most books depict Sir Isaac Newton as the man who, whilst sitting under a tree had an apple land on his head... instead of just accepting that's what apples do...fall down and land on your head.... he proceeded to ask lots of questions. He wanted to know why they moved downwards towards the earth rather than side-wards or upwards.

In his quest for answering these questions Newton became responsible for laying down the fundamental laws of the physical universe. He combined his knowledge with that of the work of the great scientific minds of his day and those that came before him; Galileo, Copernicus and Kepler, to produce a set of laws which describe not only how things work but why they work. He once confessed "If I have seen further than other men, it is because I stood on the shoulders of giants". The trouble with standing on the shoulder of giants is that if you fall off, you will experience a force far greater than the one felt if you fell off the shoulders of a dwarf!

Joking aside, the physical laws explained by Newton have allowed us to walk on the moon and build wonderful play parks such as Alton Towers and Disney World; magical places full of fun, wonder and of course physics!

These laws have also allowed us to remove some of the mystery surrounding many magicians' tricks. This is good news for us because by understanding how and why things work, we can not only repeat the tricks but improve upon them, or better still come up with our own magical tricks and mind boggling demonstrations.

Now here is a dilemma of mine which I will share with you: as a scientist very few people invite me to their birthday parties, yet... for some strange reason which escapes me, magicians are given pride of place at such gatherings. They come in – wow you, trick you, take your money and NEVER (because of a very clever marketing ploy called the Magic Circle) tell you how to reproduce their awe inspiring activities. This obviously means you are not encouraged to copy and spread the wonder of magic yourself. Whereas I, as a scientist, am so enthusiastic about the magical world that I work and live in, when given the opportunity I will try to share my enthusiasm in the hope that my listeners will take up the torch and spread the word... I don't hide the 'secrets of success'. I try to make the knowledge as plain as the open pores on your face. I tell all and sundry how they can reproduce the activities which excite and wow me. I even give health and safety training! So why am I telling you this??? I want to be invited to your parties and classrooms and to do so have decided to use the very clever marketing ploy of the users of 'slight of hand'. I have put the word Magic in the title of this exciting educational book!

Extra Bounce!

Nuts & Bolts

Football or Basketball
Tennis ball
Safety glasses

Safety: Make sure you are in an open space when you do this trick!! All assistants should wear eye protection; safety glasses, specs., or sun glasses.

Secrets for Success

Hold both balls up at shoulder height. Ask your audience what will happen when you let them go. Why do they not stay suspended in the air? Which ball will hit the ground first? Which ball will rebound with the greatest height?

Hopefully they all mention the balls drop because of gravity. The name we give to the attractive force the earth exerts on all objects on its surface and within its gravitational field. The balls will hit the ground at the same time if dropped from the same height and will rebound roughly to the same height. Did the balls rebound to the same height they started from though? No. When released the potential energy is changed into moving energy (kinetic energy). If the collision was elastic, that means no energy is lost in the colliding process, then the balls would rebound to their starting height, but in reality no collision is elastic. All collisions result in some of the energy being converted into another form of energy; the two main forms in this trick are sound energy and heat energy (due to friction).

Now hold the tennis ball on top of the football. What will happen when they are both dropped in this position? Drop the balls at the same time. Wow!! The tennis balls flies off like a rocket. Fast and high!

Repeat this trick but this time get your audience to focus on the larger football – they will see that it hardly bounces at all. Make sure you practice this trick lots of times before you do it in front of an audience. A very small variation from the top of the football can make the tennis ball fly off into the audience! An excellent example of chaotic behaviour! How could you change this demonstration to reduce the chaos aspect of the many directions that the tennis ball can fly off in?

Science in a Nutshell

This demonstration is all about the conservation of energy and momentum. Firstly the conservation of energy law states that energy can not be created or destroyed – only changed from one form to another. There are nine forms of energy: heat, electrical, light, potential (elastic and gravitational), magnetic, sound, movement (kinetic), chemical and nuclear. When we switch on the television electrical energy is changed into light, sound and heat energy. We call objects and devices which change energy from one form to another, transducers – we are transducers! We change the chemical energy in foods and drinks into energies such as heat, sound, movement and electrical (in our nerves).

All moving objects possess a quality which physicists call momentum. This quality is a means to measure the inertia of a body (the stubbornness of a body to start or stop moving or change direction). We calculate momentum by multiplying the mass of the object by its velocity – velocity is the measure of the speed of an object in a straight line or specified direction.

As with energy, momentum is also conserved. This means that the sum of the momentum held by the object or objects before the collision will equal the sum of the momentum of the objects after the collision. Snooker and football players are brilliant experimental physicists and mathematicians – they know all about collision theory, angles and spin! My good friend James Soper often says "David Beckham is a scientific genius!" and after listening to his explanation of how footballers use the laws of Newton, Galileo and Coanda to move a football in such a fashion that it not only moves majestically over the heads of the defending wall but then drops whilst at the same time follows its bended path into the back of the net you can't help but get carried away by the emotions experienced when the commentator shouts the much awaited words of "GOOAALL!!!!"

When dropped separately it is easy to see that the football will have a larger momentum than the tennis ball. They both fall at the same rate because the accelerating force (g = 10 N/Kg) acting on them is constant, but the football is much more massive. As a consequence it will have much more kinetic energy - **(kinetic energy = ½ mv²).**

So when the balls are dropped one on top of the other, after impact, the tennis ball not only leaves with its own kinetic energy and momentum but also gets most of the football's kinetic energy and momentum. So it therefore bounces 'magically and majestically' to a much higher height with greater speed.

Going Further

Because scientists don't particularly like wasting time writing – they like doing! – they have devised their own little language a bit like the short-hand script used by secretaries. The scientists' short-hand consists of letters and symbols which represent certain quantities, for example **m** is the short hand used to represent the quantity mass, **v** is used for the quantity velocity and **p** represents the quantity we call momentum. These symbols are recognised all over the world which thus allows a lot of collaborative work to take place between scientists who speak different languages. Mathematics is a universal language which when studied allows you to see the world in a completely different light. It is a tool which permits you to access the magical multidimensional world of scientific discovery.

So instead of writing 'momentum equals mass times velocity' we write
p = m x v

Much easier don't you think!

Straw Through a Potato?

Nuts & Bolts

Stiff straw
Potato
Gardening glove – for protection!

Secrets for Success

Challenge a member of your audience to stab a straw through a potato without bending or breaking the straw. Most that attempt this will be unsuccessful and thus wowed when shown the secret for success.

Place the gardening glove on your non-writing hand and pick up the potato so that the base of the potato is above the palm of your hand. Grab the straw with your stronger hand, hold it in a way that you can stab the potato whilst also ensuring that there is enough of the straw to fully pass through the vegetable and poke out the other side. Hold both the straw and the potato firmly and with a quick, sharp jab, stab the straw into and out of a side section of the potato. The glove is mainly for beginners to ensure they don't poke the straw through their hand!!

The straw must be jabbed at right angles to the potato otherwise it will buckle. If you can't get hold of a stiff straw the trick may still be done but this time you will need to place your thumb over the end of the straw – trapping air between your thumb and the potato.

How can a straw be so strong?

Science in a Nutshell

By putting your thumb over the end of the straw, the trapped air inside the straw becomes compressed once the straw enters the potato. This means the pressure inside the straw is greater than the pressure outside thus strengthening the straw; the compressed air trapped in the straw makes it become less flexible and less likely to bend during the stabbing process. However if your straw is stiff and your stabbing motion quick it will pass through the potato without the added help of air pressure! Don't have a potato? Don't worry apples and pears work just as well.

Straws are made from very thin sheets of plastic which means the edges are sharp. If you were to punch a sheet of polystyrene with the same stabbing motion (and no straw) your hand would make a dent but would most probably not pass all the way through the polystyrene. However, if you repeat the exercise with a straw in your hand the straw will pass through. You are using the same arm action and therefore the same force so why the difference? Obviously the contact surface area is significantly reduced when the straw, as opposed to your hand, is used.

Pressure (**P**) is a measure of the amount of force (**F**) acting on a certain area (**A**). And using the shorthand scientific language the relationship between these three quantities can be written as:

P = F/A

If the same force pushes on a larger area it is less concentrated, so we say the pressure is less or another more imaginative way of thinking about this is to envisage the following two scenarios. Firstly your local fireman wearing very sensible flat boots has just jumped off of his ladders onto your foot. Ouch says you. Secondly the same fireman wearing stiletto heels (Mmm!?) jumps onto your foot. How much larger would your cry of pain be?? His weight force remains the same; the only parameter we have changed is the contact area of his shoe. Even though it would be quite amusing to see the second scenario the pain level would be increased at least 10 fold so I think I will stick to my calendar fantasies rather than carry out the experiment!

In fact the pressure is said to be inversely proportional to the area. That means if the area is halved the pressure is doubled and so on. Thus going back to the polystyrene example, the straw carries a 'punch' roughly 300 times larger than that of the fist! It is for this reason that until you are confident of your aim you must wear a gardening glove to protect your hand in case you miss the potato!

Fact

How do drinking straws work?

When we put a straw into a drink the air pressure acts on us, the top of the drink and on the fluid in the straw. When we suck we use our mouth to reduce the pressure over the fluid, inside the straw – the sucking process removes some of the air molecules inside the straw, this decrease in the number of molecules means the number of collisions acting downwards on the fluid as a result of the air is decreased; the air pressure is lowered. The pressure over us and the rest of the drink however remains the same. Due to the fact that there is now a pressure difference there will be a movement of fluid from the high pressure region to the low pressure region, thus the fluid is pushed up the straw into our mouths.

Useful tip

Flowers too short for the vase? Add extra length to them by inserting the stems into drinking straws. This also works to prop up sagging flowers.

Air Presses Up!

Nuts & Bolts

A drinking glass
Water
Measuring jug
Piece of stiff card – bar mats are perfect
Graph paper
Assistant

Secrets for Success

Fill the glass to the top with water. Slide the piece of stiff card over the top of the glass. Hold the card tight against the glass and turn upside down over your assistants head! Take away the hand holding the card. Voila!

How does the water stay in the glass?

Science in a Nutshell

The water in the beaker has a weight force acting down due to gravity. For the water to remain in the beaker there must be a force, larger than the weight of the water, acting upwards. Where does this force come from?? The air surrounding us!

The air is made up of billions of gas molecules. The main ones being nitrogen ~ 80 % and oxygen ~ 20 %. These molecules bump into us all the time. Each collision is like a small punch - the size of the force due to one molecule hitting us would be very difficult to measure because it is so small – but we are being bombarded by billions of the molecules every second of every day, and thus we experience a force of 10 Newton's (N) acting on every 1 cm^2 of our skin all the time. This is very measurable! To feel the effect of a 10 N force pick up a 1 kg bag of sugar. Heavy! So why aren't we crushed??

This trick shows us that the air pressure acts in all directions; up, down and sideways. This is good news for us because if it only acted in a downwards direction we would be a race of pancake people!

Going further

Place the dry glass upside down on a piece of graph paper and draw around the lip of the glass. Remove the glass – count the number of 1 cm^2 squares enclosed within the circle. Let's say you calculated an area of 10 cm^2. This means there is a column of air above this circle with a mass of 10 kg pushing downwards with a force of 100N! Notice that to

convert masses to weight forces the mass value in kilograms is multiplied by the number 10 – very convenient.

Weight = mass x gravity

W = mg; where g = 10 N/kg

10 N/kg means that a 10 N force is exerted on every kilogram of mass at the Earth's surface. (Actually it is a little less than this, but for ease of calculations we round the value up to 10). Weight is a force and is

measured in Newtons, mass is a measure of the amount of 'stuff' in a body. It doesn't matter where I go in this universe I will always have 70 kg of blood, guts, fat and bones, but the forces pulling on me will vary depending on whether I am in deep space; where I will be weightless, on the moon; where gravity is 1/6th of that on earth or on a planet 200 times larger than earth whose gravitational forces of attraction would crush me.

Air pressure acts in all directions so when the card is underneath the up-turned glass there is a force of 100 N acting upwards on the area of 10 cm^2 due to air pressure and a force acting downwards due to the weight of the water contained in the glass.

Pour the water into the measuring jug. 10 ml of water has a mass of 1 g thus 1 litre of water has a mass of 1 kg – test this fact using measuring scales. You will find that the mass of 1 litre of water is in fact 1 kg. Nice and convenient, magical in fact.

Let's say you have a glass that can hold a volume of 300 ml of water. This volume will have a mass of 300 g. Mass is just a measure of the amount of stuff (atoms and molecules) that are in the object under investigation. To calculate the weight force that this mass of water pushes down with we have to convert our masses in grammes to kilograms (kg) and then multiply this value by 10 to give us the force in Newton's (N). So 300g is the same as 0.3 kg. Therefore 300 ml of water has a weight of 3 N (0.3 x 10 = 3). Remember 'weight = mass x gravity'.

Confused??!! Don't worry this conversion stuff confuses many people – the main reason for this being the case is that in every day life when we talk about how much 'stuff' there is in an object we talk about it's weight when in the world of science – which we all live in by the way, we should say it's mass. "My bananas have a mass of 2 kg, this means their weight is 20 Newtons." When I jump on the bathroom scales I 'weigh' myself (and usually get depressed!) and find that there is 70+ kg of blood, guts, hair, teeth, undigested food etc. in my body. I would have this mass of stuff no matter where I was in the universe, on the moon, outer space or on one of the exotic moons of Saturn. On earth I am pulled towards the centre of the planet because of the gravitational attractive force between me and the planet. At the earth's surface this pull has a force of roughly 10 N for every 1 kilogram that it pulls on. So I therefore have a weight of 750 N. Oops I mean a weight of approximately 700 N!

Going back to the water in the glass – theoretically you could change the trick so that the weight of the water acting down just balances or is slightly greater than the air pressure acting up. How could you do this? To answer this problem you will have to think about decreasing contact area and increasing the height of the vessel.

Wood You Believe It!

Nuts & Bolts

**Strip of balsa wood
3-5 mm thick
Sheet of newspaper
Graph paper
Pen
Baseball bat
Table**

Secrets for Success

Place the strip of wood on a flat table with a quarter of it hanging over the edge of the table. Place the sheet of newspaper over the strip. Run your hand over the paper to flatten it and remove any air trapped beneath it. Ask your audience what they think will happen to the paper when your

volunteer brings the bat quickly down on to the strip over hanging the edge of the table. What do you think will happen?? THWACK! Wood snaps and paper remains on the table.

Why didn't the paper move?

Science in a Nutshell

The huge column or air above the newspaper has a weight of almost 10 N (1 kg) per cm^2 pushing the paper onto the table. Use a ruler to find the area of the sheet of newspaper in cm^2 and then multiply that area by 10 to give you the size of the force needed to make the paper fly through the air. A very large force yes?! So unless you are a super human it will be pretty nigh impossible for you to lift the paper. In fact if you hit the wood quickly enough the wood will break near the edge of the table – very noisy and impressive!

Going Further

Tear the paper in half thus reducing its mass and the area in contact with the air column by half, place it on top of the strip and hit away! How many times do you need to go through this process until the paper flies?

Put your hand on a sheet of graph paper. Draw around the edge and roughly work out the area of your hand in cm^2. The area of my hand is 54 cm^2 which means when I put my hand out I hold up 54 kg of air! How much air can you hold up? Bet you didn't realise you were that strong!! What is the smallest area of paper you can use before you hit and cause the paper to fly?

There is a huge pressure due to the surrounding air acting on us, pushing in on us, every moment of every day. Why are we not crushed?

Our bodies are filled with liquids and gases which push outwards against our skin. The outward pushes are balanced by the inward pushes because the pressures are equal in size but act in opposite directions. What would happen if the infamous Scotty made a mistake and beamed Dr Spock up into outer space? POP! He would blow up like a balloon and explode. YUK!

Do Not Open!

Nuts & Bolts

An empty 1 litre plastic bottle with lid
Drawing pin
Water
Permanent marker pen

Secrets for Success

Use the marker pen to write 'DO NOT OPEN!' in large letters on the lower section of the bottle. Use the sharp drawing pin to poke several tiny holes through the bottle at the bottom of the letters – the pen colour will hopefully not allow your unwitting prey to see the holes! Over the sink fill the bottle to the very top. Screw the lid on with the tap still running. Water will escape through the holes during this process – don't worry, all will be revealed! With the lid on tight, lift the bottle by the lid and place the bottle on a table or work-surface with the words in full view to anyone entering the space. Step back and casually watch, even play dumb if someone asks you about the bottle – eventually someone will open it – and then wish they hadn't!

Why didn't the water leak out of the bottle whilst the lid was on?

Science in a Nutshell

When the lid is on air pressure can't push down onto the surface of the water. The air molecules can bombard the lid but are prevented from hitting the surface of the water by the lid. The air molecules act in all directions so will also be pushing against the whole outside surface of the bottle. This push keeps the water in the bottle. However when the lid is removed the air molecules can now push down on the surface. The atmospheric pressure caused by these molecules along with the downward force from gravity team up. These pushing and pulling forces ensure that the water pours out of the holes soaking the nosey, instruction-disobeying person. Let's hope it's not anybody heading out to a very important meeting!

Going Further

Our earth is surrounded by a layer of air that is 483 km deep. The layer is thickest at the earth's surface because the gravitational forces of attraction are largest here. As you move further away from the centre of the earth the gravitational force of attraction gets smaller. This means that the force of gravity is smaller at the top of Mount Everest than it is at sea level. If there is less force pulling you down at high altitudes does this mean you can jump higher??

As a consequence of decreasing force the thickness of the atmosphere decreases, the force of gravity is not strong enough to hold a thicker atmosphere. This means that water will boil at lower temperatures at the top of very high mountains because less air molecules are pushing down onto the surface of the water. This is not so good for you tea lovers – and, anyone in a rush for their cooked dinner will have to wait because the lower temperatures mean it will take longer for the food to cook!

Fact

New Heights

When Mexico City was awarded the 1968 Olympic Games, many low-lying countries protested, citing concerns over the host country's high altitude and extreme climate. At the city's altitude of 2300 m, the air contains 30 per cent less oxygen than at sea level! These fears had foundation. The winning times of the 5000 m, 3000 m, steeplechase and 10000 m were the slowest in 20 years. On the plus side the gravitational force pulling the Olympians to the earth was slightly 'relaxed'. Bob Beamon's giant leap of 8.90 m, a full 55 cm further than the old record wowed the world. This record was finally beaten in 1991.

Vacuum Pack Your Granny!

Nuts & Bolts

**Large bin liner, preferably 70 litres +
Vacuum cleaner
An assistant**

Safety: Never place the bag over anyone's head or have plastic bags in the reach of small children.

Secrets for Success

Ask your assistant (minus shoes) to carefully step into the bag. If your assistant is a large adult they may have to crouch down in the bag as

I want you to be able to secure it around the base of their neck! Give the nozzle of the vacuum cleaner to your assistant asking them to cup a hand around the end to prevent the bag being sucked into it and blocking it once the cleaner has been switched on. Gather the bag around the hose and their neck – obviously don't hold it too tightly! Switch on the vacuum cleaner. The bag will shrink tightly around their body – the result is hilarious!

Science in a Nutshell

Before the vacuum cleaner was switched on the bin bag contained the assistant and air. At this time the air pressure inside the bag equals the air pressure outside of it. When switched on the fan inside the cleaner spins rapidly causing an area of low pressure to be formed inside the cleaner. Air always moves from regions of higher pressure (inside bag with your assistant) to regions of lower pressure (inside the vacuum cleaner), hence the air in the bag quickly moves up the tube to replace the air lost in the cleaner. If you don't allow air to re-enter the bag to replace that which is lost, the air pressure in the bag will decrease.

There will now be a difference between the pressures inside and outside of the bag. The effect of this pressure difference is felt by the assistant and viewed by your audience! The bag is pushed onto the assistant and is molded around their body! Switch the cleaner off as soon as this effect is seen – do not be surprised if everyone in the room wants to be vacuum packed as well!

Going Further

There are many practical applications using the science behind this trick. We vacuum pack some foods, to not only increase their shelf life, but to also reduce spoilage. If you wanted to store your duvets for the winter you can place them in a bin bag and use the above science to remove all of the excess air – you will be surprised at how small they become once all of the air has been removed from between the fibres. You could of course use the same method to allow you to take more clothes on holiday with you – unfortunately though you are not guaranteed a vacuum cleaner to re pack for the journey home!

Fact

Vacuum-packing

Wayne Goates was the first scientist to vacuum-pack one of his volunteers. As a result of this you can now commercially buy vacuum seal storage packs. These packs seal your clothes and help you gain up to 300 % more storage space by removing the air between the fibres of the cloth!

The Incredible Growing Marshmallows

1 The Mallow Expander

Nuts & Bolts

An empty clear glass wine bottle
1 bag of small marshmallows
Wine vacuum pump (available from large department stores or off-sales)
Large dark elastic band

Secrets for Success

1/3 fill the dry bottle with marshmallows. Push the elastic band down over the neck of the bottle until it is level with the top of the marshmallows. Push in the grey plastic cork. Attach the pump, as given by instructions, and start pumping. 100 – 200 pumps, depending on the size of the bottle should give a significant growth! Shake the bottle every 20 pumps to keep the marshmallows free. Remove the pump and note how much the mallows have expanded by comparing the new height to the height of the elastic band. Hold the bottle up to your audience and quickly pinch the cork plug open. The marshmallows will shrink back to their original size nearly instantaneously – always gets a wow!

Kitchen vacuum packers do the same thing even more dramatically but are not so easy to purchase. Marshmallow bunnies can be turned into monsters. Very funny!

Science in a Nutshell

The job of the vacuum pump is to remove the air above the wine, thus creating a partial vacuum. The oxygen in the air can react with the wine. It oxidizes it and makes the taste of the wine less tasty! The marshmallows allow us to see this process in action. They are filled with air and in normal circumstances the outward push of the air in the marshmallows is equalled by the inward push caused by the air molecules surrounding them, a bit like a tug of war between two teams of equal strength. Removing the surrounding air molecules decreases the surrounding air pressure and the air inside the marshmallows expands to take its place. It is this outward expansion that causes the marshmallows to grow – they shrink again when the air rushes back into the bottle.

2 The Mallow Masher!

Nuts & Bolts

1 bag of small marshmallows
Empty plastic drinks bottle
Commercial fizz-keeper: pressurising pump

Secrets for Success

Half fill the bottle with marshmallows. Screw on the pressurising pump. Begin pumping and notice what happens to the marshmallows. They shrink to a third of their original size. Keep looking at the marshmallows. Release the pressure by unscrewing the cap. The marshmallows 'magically' grow again.

Science in a Nutshell

The fizz-keeper is similar to a bicycle pump. With each pumping action air molecules are forced into the bottle. This increases the pressure inside the bottle. The air pressure surrounding the marshmallows becomes much greater than the air pressure inside the marshmallows. Each marshmallow is bombarded by many millions of added molecules – these extra collisions make the air in the marshmallow contract – the volume decreases – and they shrivel up.

The Collapsing Can

Nuts & Bolts

An empty aluminium soft-drinks can
Large bowl of cold water
A pair of kitchen tongs or tongs large enough to hold the can
A heat source

Safety: Do not heat the can over high heat or heat the can when it is empty. This may cause the ink on the can to burn or the aluminium to melt.

Secrets for Success

Fill the bowl with cold water. Put 15 ml (1 tablespoon) of water into the empty can. Bring the water to the boil by heating the can on a stove, or by holding it over a propane camping burner with the aid of the tongs. When the water boils, a cloud of condensed vapour will escape from the opening in the can. Allow the water to boil for about 30 seconds. With the aid of the tongs, grasp the can and quickly invert it and dip it into the cold water in the bowl. The can will crush violently!

Science in a Nutshell

Before you heat the can, the pressure inside the can, due to the air molecules hitting the inside surface, is equal to the pressure outside the can, which is also due to air molecules hitting the outside surface.

The water molecules in the bottom of the can are moving about bouncing off each other like billiard balls. The average speed of these molecules is characteristic of the state of the substance: it gives the temperature! A substance when hot has faster moving molecules than when cold. So a solid, such as ice, at a certain low temperature has its molecules vibrating at low average speeds. They are not moving about, just vibrating in what physicists call a 'potential well', but if we increase the temperature the vibrating speeds increase until eventually the molecules start escaping this 'well'; they have enough energy to overcome the strong 'sticky' attractive forces holding them in a solid state, the molecules will move off on their own. This is known by the highly technical term of 'melting' and for solid water this occurs at 0 ºC.

The heat source will heat up the water in the can, that is the water molecules will gain more energy and their average speed will increase; when we bring any liquid to the boil we give its molecules so much movement energy that the speeds are increased to a level which will eventually allow them to escape the small attractive 'sticky' forces which keep them in a liquid state. They will then form a gas or a vapour, for water this occurs at 100 ºC.

The vapour from the boiling water expands moving in all directions and pushes the air out of the can. By continuing to boil the water for 30 seconds after it has reached boiling point we are hopefully removing all of the air molecules that were originally in the can. The can is then just filled with water vapour; you cool it suddenly by inverting it in the bowl of cool water. This action will cause the water vapour to cool and condense (turn back into water) - a partial vacuum is created, the air pressure inside the can is now very much lower than the outside air pressure. This pressure difference between the atmospheric pressure and the vacuum is responsible for the rapid crushing of the can.

You may expect the water in the bowl to fill the can through the hole in its top; some water may do this, however the water cannot flow quickly enough into the can to fill it as the air pressure acts immediately!

The Tablecloth Trick!

This is a classical trick and if it works you will leave everyone spell bound and if it doesn't it will leave them initially in shock and then in a heap of laughter! Either way Fun is Guaranteed!

Nuts & Bolts

Tablecloth without hems
Flat smooth table
Selection of heavy crockery with smooth bases

Secrets for Success

The aim of this trick is to yank the tablecloth from under a china place setting without destroying the crockery!

Spread the tablecloth onto the tabletop. Initially, for beginners, make sure the far side of the cloth does not hang over the back of the table. Have at least 50 cm of cloth between you and the edge of the table. Put your choice of crockery in the middle of the table, not too close to the edge that you are pulling towards! Grab the ends of the cloth and bring your knuckles to the edge of the table. There should now be a dip of about 25 cm in the cloth between you and the table. With both hands pull the table cloth away from the table – with no hesitations – the pull must be towards you and parallel to the table; if you pull upwards that is the direction in which the crockery will fly!

Start small, with one plate over a carpeted or cushion covered floor, then gradually work up to a full table setting.

Science in a Nutshell

Newton is most famous for his laws of motion; his first Law of Motion which is also sometimes called 'the Law of Inertia', is responsible for the current police laws on the use of seat belts in cars – it's interesting to note that it only took us 250 years after he came up with his law for us to introduce ours!

The secret science behind this trick is based on Newton's First Law of Motion which states that bodies in motion tend to stay in motion and bodies at rest tend to stay at rest unless acted upon by an external force. The word inertia is linked to how heavy and how fast the moving object is; the more massive or heavy the object is or the faster it is moving the greater is its resistance to a change in its state of motion. This is why we use heavy crockery it has more inertia and a greater resistance to a change in its state; which is initially stationary!

When you pull the cloth, frictional forces act on the objects in the direction of the pull – because the cloth and the bases of the crockery are smooth and slippery, these frictional forces are small. When the cloth is free of the objects they are left in contact with the table – here the frictional forces are greater. The objects slow down (decelerate) very quickly and come to rest in a short distance – hopefully the distance is so small no one will have noticed that they moved at all! The faster you pull, the less distance they will move.

We experience this phenomenon when we ride in a car or on a bus. When moving, both you and the vehicle move at the same rate; if the brakes are applied the brake pads push against the tyres and the contact frictional forces cause the vehicle to slow down. If the pressure on the brakes is gradual the vehicle slows down slowly and the frictional forces between you and the vehicle (either you and the seat, if seated or you and the floor, if standing) help to slow you down slowly. But if an emergency stop is required, the contact forces between you and the vehicle will be too small to slow you down and according to Newton's Laws of Motion you will keep moving in a straight line until a large enough force is experienced by you to stop you. This is the job of a seatbelt; if not worn then it will be job of the windscreen or the seat/person in front of you and remember if the vehicle is moving at 50 km/hr then so are you!

If you are standing on a bus and it suddenly moves forwards you will experience a sensation of moving backwards. In fact, as Newton stated: stationary objects want to stay stationary and the heavier they are the more stubborn they are to being moved, so, you try to stay in the space you were when it started moving. Unfortunately, our feet are in contact with the floor of the bus and the frictional forces due to this contact pull our feet forwards with the bus. Due to the fact that our bones and skin make us a whole being eventually the rest of our body follows our feet. But there is a measurable time delay. It is this time delay that we use in the tablecloth trick. The cloth is moved before the heavy crockery picks up enough momentum to start moving.

Egg Drop Shock!

Nuts & Bolts

4 clear plastic beakers half filled with water
4 eggs (or golf balls for the less adventurous!)
4 tubes – insides of toilet rolls or empty vitamin C tubes
1 baking tray
Table top
1 garden twig brush (the witchy type!)

Secrets for Success

Place the four beakers on the table 2 x 2. Place two close to the edge of
the table and the other two behind. The spacing is dictated by the size

of your baking tray. Place the baking tray on top and then place the 4 tubes on the tray such that they are placed above the centre of the cups. These are the 4 wheels, the body of the bus as well as a driver and three passengers. Imagination is a wonderful gift! Ask your audience what is missing – they will immediately say 'HEADS!' You don't want to disappoint so out come the eggs – decorated with faces if you so wish. Place the eggs carefully on top of the tubes.

This is where the fun starts. Remember Newton said stationary objects want to stay stationary and objects which have a force acting on them will move in the direction of that force. Well you are going to hit the tray making it travel forwards. The path it takes will not be a straight line but curved because gravity is also acting in a downwards direction upon it. The tubes are in contact with the tray and will fall away from under the eggs. No forward force is exerted on the eggs. So under the influence of gravity they will fall downwards into the cups!!!! I have done this trick a hundred times and I still get the wow factor seeing the eggs in the beakers.

The secret weapon is the instrument used to hit the tray – the gardening brush. If you put your foot on the twigs and pull the handle towards you, you ensure lots of potential energy is stored in the brush. No work is done by the brush until you let go of the handle. When you do, the handle shoots forwards very quickly. The edge of the tray should be at least 2 cm over the edge of the table. The Brush is placed at the base so that when the handle is released it will hit the tray. You will need to stand at right angles to the table to do this. Your right foot should be standing on the twigs – it is the pivot point about which the brush moves. When released the brush must hit the middle of the tray. Pull the handle back through a distance of 20 cm – no more. Get your audience to count you down. 3 2 1 THWACK! All eggs will magically end up in the cups! Amazing!

It is wise whilst practicing this trick to use golf balls or hard boiled eggs – for obvious reasons!

Science in a Nutshell

So, basically inertia means the amount of effort needed to get something to start moving- or the effort required to stop a moving object. However, Newton's Laws do not mention anything about measuring the effects of motion.

A moving object has two characteristics: its mass (m) and its velocity (v). These result in the measurement of the object's momentum and kinetic energy. This brings us back to two other laws concerning motion; the Conservation of Momentum and the Conservation of Energy.

Momentum is a way to measure inertia, and to calculate the momentum that a moving object has, we need to multiply its velocity (v) by its mass (m). Using our scientific shorthand the formula *momentum (p) equals mass times velocity* becomes $p = mv$ (much shorter). Have you ever heard of a sports team having momentum? This means that the team is playing so well, that it would be difficult to stop them. This is the same for heavy objects moving quickly.

The Law of the Conservation of Momentum states that when one object hits another, within a closed system (that means no other forces are involved), the total momentum of the objects after the collision is equal to the total momentum of the objects before they collided.

A moving object not only has momentum but it also has kinetic energy (KE). This energy is proportional to the mass (m) of the object [as mass doubles KE doubles and vice versa] and the square of the object's velocity (v) [if the velocity is increased by 3 the KE is increased by the square of 3 =9]:

KE = ½ mv²

The law of the Conservation of Energy states that the total amount of energy in a closed system is constant. In other words, if we have a system where no matter (atoms, molecules, particles etc.) or energy enters or leaves, the amount of energy available remains the same. The energy may convert from one type to another, such as kinetic to sound and heat, as long as the total amount is constant. We will explore this idea further on in the book.

Newton's Second Law quite simply states that the more force (F) applied to an object of constant mass (m) the faster it will accelerate (a) or the more quickly it will decelerate (-a), obviously depending on the direction the force is applied with respect to the direction of motion. Mathematically the law links the change in momentum of the moving object with the time taken for that change to occur. *The resultant force equals the rate of change of momentum* or more simply F = ma

Fact:

Crumple Zones

If a fast moving, heavy car is brought to rest in a very short period of time the change in momentum will be very large hence the forces acting on the passengers will also be very large. The force acting is given by Newton's 2nd Law

$$Force = mass \times \frac{\Delta \ velocity}{\Delta \ time}$$ (Δ is the scientists shorthand for 'change in')

Since 1966 car manufacturers have built their cars with crumple zones at the front of them. Engineers place weak spots in strategic locations which allow the metal work of the car to collapse in a controlled manner if it is involved in a crash. Since it takes time for the metal work to collapse the stopping time is increased, thus the force acting on the passengers is greatly reduced. The crumple zones also remove energy from the impact by (1) using energy up by converting it to heat in the process of deforming the metal work and (2) directing the energy away from the passenger area and channelling it to areas such as the floor, roof and bonnet.

Newton's Third Law describes how objects interact with one another. When a boxer punches somebody in the face; the face pushes back on the hand with an equal force but one which acts in the opposite direction – the result being both the hand and face will experience pain!! Obviously this will not be one of the experiments explored in this book.

So basically if a pulling or pushing force is exerted on an object and that object remains at rest it will exert an equal force which acts in the opposite direction to the initial force. When I stand on the ground, because of gravity, I exert a force of 700 Newton's on the earth. The earth pushes back on me with an equal force, a reaction force, in an upwards direction. If these forces were not equal I would be moving. So if the upward pushing force of the earth was removed I would fall downwards towards the centre of the earth just like Newton's apple. Likewise if whilst driving, my car skids and hits a wall the wall will impart on my car the same force with which it hit it! A very big ouch if there was no crumple zones!

Useful tip

Keep car windows frost free by coating the windows the night before with a solution of three parts vinegar to one part water.

Chapter 2
Chemical Opposites

Acids and Bases

The word acid comes from a Latin word meaning sharp or biting to the taste. You would have experienced this sensation if you have ever sucked a lemon! Many acids occur naturally and, in the kitchen, the two most common are citric acid (found in citrus fruits) and ethanoic acid (vinegar – a dilute solution of ethanoic acid formally known as acetic acid). Most of the acids found in and around the home are weak but some are very strong, they are poisonous and extremely dangerous. One of the strongest acids is hydrochloric acid, HCl, a compound made up of two atoms; one hydrogen atom and one chlorine atom. This acid is used for soldering but we also find it in our stomachs which is why, if you have a stomach ulcer, you suffer so much pain. Sulphuric acid is another strong acid which is found in the battery of cars. They most definitely should not be tasted!!

The chemical opposites of acids are bases. Bases are usually found as a solid compound and for it to react with an acid it is usually dissolved in a solvent, such as water, to form a solution. Bases in solution are called alkalis. At home the most common place to find bases and alkalis is under the kitchen sink! They are very good cleaners; examples would be washing soda crystals and bleach. You will also find some bases in the food cupboard; baking soda (sodium bicarbonate) being the most common.

When a base (or alkali) is added to an acid, it will neutralise the acid's properties and vice versa.

In this chapter we are going to investigate acid-base reactions using vinegar (ethanoic acid CH_3COOH) and baking soda (sodium bicarbonate $NaHCO_3$). When these two chemicals are mixed a salt ($NaC_2H_3O_2$), water (H_2O) and carbon dioxide (CO_2) are produced. Water is a liquid (l), carbon dioxide is a gas (g). If water is used to dissolve a solid (s) to form a solution the scientific symbol for aqueous (aq) is put after the chemical. Hence the reaction can be written as:

$$CH_3COOH \ (aq) + NaHCO_3 \ (s) \rightarrow NaC_2H_3O_2 \ (aq) + H_2O \ (l) + CO_2 \ (g)$$

By using these symbols scientist cut down on a lot of writing time which ultimately gives them more time to experiment, think and solve problems. By mastering the scientific language a whole new world is opened up for you to play in – just like learning Spanish will allow you to enjoy the company of the Spanish speaking communities as well as appreciate their culture and history. More books to read, more songs to sing!

So let's make some carbon dioxide.

Balloon Blow Up

Nuts & Bolts

Vinegar
Small clean plastic bottle
Baking soda
Small funnel
2 balloons
2 sets of kitchen scales
Tall glass beaker
Small candle and matches
Assistant
Safety glasses

Safety: Protective eye wear should be used in case the balloon bursts.

Secrets for Success

Put two tablespoons of baking soda into one of the balloons, then third fill the plastic bottle with vinegar. Very carefully stretch the opening of the balloon over the neck of the bottle without allowing any of the powder to enter the bottle. Ask your assistant to hold the bottle in their left hand, preferably with their hand in contact with the neck of the balloon and bottle - they are to use their right hand to gently lift the section of the balloon holding the powder and allow the powder to drop into the bottle. At this stage lots of carbon dioxide gas will be formed and the reason for holding the balloon onto the bottle will become apparent! If the balloon isn't held firmly enough it may pop off, but hopefully it will remain in place getting bigger and bigger – much to the concern of your assistant and joy of your audience.

When the balloon is more than ¾ blown up remove it from the bottle and tie the neck.

Blow up the second balloon so that it is roughly the same size as the one containing carbon dioxide. Tie the neck. Place the balloon in your assistants' hand. Ask them how much it weighs. Now place the balloon containing the carbon dioxide on their other hand, do they experience a difference in weights? (They should!) Place the balloons on the kitchen scales. Is there is measurable difference in weight?

Ask your assistant to hold both balloons up and drop them together – which one hits the ground first? They should in theory hit the ground at the same time because the same accelerating force – gravity- is acting on them. In fact the one filled with CO_2 hits the ground first, this is because as well as the force of gravity acting downwards there is also an upward force acting on the balloons due to friction caused by collisions with the surrounding air. The effect is similar to that of a feather being dropped. The lighter balloon, the one filled with air, feels the effects of the upward air resistance more than the other heavier balloon.

Pour some vinegar into the glass. Add a large tablespoon of baking soda to it. Put the glass down onto the table top. Light your candle. Lift up the glass and pour the heavy carbon dioxide gas over the flame. The flame will be extinguished.

Science in a Nutshell

When the vinegar (ethanoic acid) is mixed with the baking soda (sodium bicarbonate) a chemical reaction takes place. Lots of carbon dioxide bubbles are formed in a very short period of time.

This experiment is an example of a reaction between an acid (vinegar) and a base (baking soda). Such reactions typically form a "salt" and water.

ACID + BASE → SALT + WATER

Because the acid component in this experiment is ethanoic acid, it allows the production of one of the products to be sodium ethanoate. That is the stuff referred to as the "salt." In this experiment the base has a carbonate component; hence carbon dioxide is also formed.

The fancy shorthand symbols used by scientists to represent our reaction are:

CH_3COOH + $NaHCO_3$ → $NaC_2H_3O_2$ + H_2CO_3

Ethanoic acid   sodium bicarbonate   makes   sodium ethanoate + carbonic acid.
(vinegar)   (baking soda)

The $NaC_2H_3O_2$ is the salt called sodium ethanoate. The H_2CO_3 (carbonic acid) then breaks down into water and carbon dioxide:

H_2CO_3 → H_2O + CO_2

This break down is rapid and the balloon expands quickly.

Instead of a bottle you could use a medium sized Ziploc™ bag and wrap a couple of large spoonfuls of baking soda in a thin piece of tissue – this is the detonator! Get your assistant to hold the parcel of powder on the inside of the bag above the vinegar level whilst you securely fasten the bag. At this stage you can bring out the safety glasses and possibly a builders safety helmet (just for effect! Watch your assistants eyes grow with concern!) Step back and ask your assistant to release the detonator!

The baking soda will start to react with the vinegar. Lots of gas will be formed as a result and the bag will get bigger and bigger until...POP! The pressure inside becomes too great and the seal is broken showering whoever is holding the bag with its contents!!! Oh and I should point out that it has been known for the assistant to panic, drop the bag and run!!! So a drip tray is recommended!

Carbon dioxide in the liquid form is found in many fire extinguishers. For it to be liquid it is kept at certain pressures and temperatures. An extinguisher can only be used once for this reason. When released the liquid CO_2 quickly changes into a gas. This gas is heavier than air and sinks downwards. If aimed correctly the gas forms a blanket covering over the fire – blocking out the oxygen required for combustion. The fire is extinguished.

Most of the time carbon dioxide is a gas. It is in the air around us, plants use it to make food, and we breathe it out as a waste product due to the respiration reactions taking place within the cells of our bodies. In scientific shorthand carbon dioxide is written as CO_2, this is a very useful way of letting us see what and how many atoms are used to form carbon dioxide. The C means there is one atom of carbon and the O_2 means there are two atoms of oxygen.

Solid CO_2 is called dry ice, unfortunately this is very difficult to make at home so we will not be experimenting with it in this book.

Everything is made of chemicals, and all chemicals are made of tiny particles called atoms. During a chemical reaction, one group of atoms are shuffled and taken apart, they get mixed with the other atoms to form a different group and make a new chemical. Similarly when vinegar is mixed with bicarbonate of soda (baking soda) one of the new chemicals formed is the gas CO_2. The bubbles of this gas can be used to turn a pop-lid drinks bottle into a most amazing volcano!

Going further

Get your wards to some research. When man landed on the moon one of the experiments carried out was the dropping of a feather and a heavy weight. What happened when these objects were raised up to the same height and dropped at the same time?

Get three pieces of A4 paper and three assistants. Scrunch one into a ball; give this to assistant number one. Give the other assistants a flat page; one is to hold it vertically the other horizontally. They all weigh the same which means they have the same mass and the same force of gravity pulling them downwards, so what will happen when they are dropped? This is a nice and simple way to demonstrate the effects of air resistance. The scrunched ball with the smallest contact surface falls first, then the vertical sheet, which is able to cut its way through the air and lastly the horizontal sheet which has the largest upward force acting on its surface area.

Vesuvius in a Bottle!

Nuts & Bolts

Vinegar
Washing up liquid
Baking soda
Empty fruit shoot bottle (or any other with a pop-up lid)
Small bucket

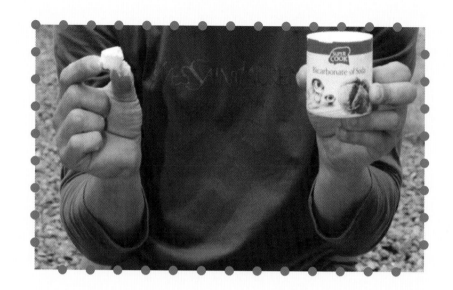

Secrets for Success

This is a demonstration for outdoors! Remove the plastic cap covering the pop-up lid and fill it up with baking soda. Set aside on the floor. Pour the contents of the juice into a glass for you to enjoy later… 1/3 fill the bottle with vinegar, add a squirt of washing-up liquid to the bottle. Ensure the pop-up lid is down. Drop the cap containing the baking soda into the bottle – this is your detonator!! Very quickly screw the lid back onto the bottle – if you are too slow the contents will pour out all over you! Place the bottle upright on the floor and step back. Watch, wait then WHOOSHH!!! The volcanic eruption will reach heights of several metres – colours could be added for greater effect.

Wash out the bottle, lid and cap and repeat the above experiment though this time place the bottle upside down in a small bucket – make sure you are in a spacious area – a field would be good! The bottle will fly off like a rocket!

I use the fruit troop bottle because the cap covering the pop-up section fits into the bottle this is not the case for the majority of drinks bottles – so if using a different bottle you will have to use your imagination to figure out a way to 'dump' in the powder so that you have enough time to put the lid on.

Safety: The bottle leaves the bucket at great speeds so care must be taken to ensure that it is not aimed at any one or anything that could be broken! An unsuccessful lift off does not mean the chemical reaction hasn't taken place; it only means not enough pressure has been produced to push up the lid – so the bottle will be pressurised!!! DO NOT UNSCREW THE LID the bottle and lid will separate at great speeds and could injure somebody. Lift up the pop-up section to release the pressure before unscrewing the lid, this will mean the contents will leave the bottle, safety glasses must be worn. Pressurised containers can be very dangerous so never leave the bottle in a pressurised state.

Science in a Nutshell

When the vinegar mixes with the baking soda, carbon dioxide gas is formed. The washing-up liquid traps the bubbles of this gas making the wonderful foamy mixture which erupts out of the bottle when the pressure inside is large enough to lift the pop-up lid. Some lids are so difficult to lift I have to use my teeth – so imagine the size of the pressure required to do the same job!! This is why the mixture goes so high and why we have to treat the pressurised bottles with respect and due care.

For the explanation of the chemistry and the effects of the build up of pressure and the resulting rocketry read the Science in a Nutshell sections of the Balloon Blow Up and 3 2 1 …. Lift Off!

mmmm... Honeycomb!

Nuts & Bolts

250 g of sugar
120 ml of water
30 ml of vinegar
30 ml of golden syrup
1 teaspoon of baking soda
A large saucepan
A baking tray
A few hungry assistants to eat the honeycomb

Secrets for Success

Mix the sugar, vinegar, golden syrup and water in a large saucepan. Bring the mixture to the boil, and then continue boiling until it turns the deep golden yellow colour of honeycomb. This takes a little time.

Remove the mixture from the stove, and stir in the baking soda. Stir thoroughly ensuring that all of the baking soda has been mixed in. The carbon dioxide produced causes the mixture to froth up, so be careful it doesn't spill. Pour the mixture into a baking tray that has been greased with butter. When it is cool, cut or break the honeycomb into pieces to eat. Mmmmmm yummy!! Why not sprinkle it over some vanilla ice cream or dip into melted chocolate?

Science in a Nutshell

The vinegar (ethanoic acid) reacts with the baking soda (sodium bicarbonate) to produce carbon dioxide gas. It is this reaction (see **Balloon Blow Up**) which puts the bubbles into your hot toffee to make this delicious honeycomb!!

Fact: Clean your dishwasher

Clean the dishwasher by running a cup of vinegar through the whole cycle once a month to reduce soap build up on the inner mechanisms and on glassware.

3 2 1 ....Lift Off!

Nuts & Bolts

Film canister with a snug-fitting lid
Some water
Blu-tac
Stop watch
Lots of old newspapers – to cover floor and tables
Measuring cylinders (5 ml or 10 ml)
Tissue paper
Blu-tac
Vitamin C tablets or Alka-seltza tablets
Stop watches
Lots of assistants

Safety: Once turned over tell your science investigators that no one is to place their heads over the top of the canisters as they take off at great speeds and could injure their faces – if their rockets don't explode they are to approach the canister from the side and carefully separate the lid from the container and then start again. This experiment is best carried out outside.

Secrets for Success

This activity needs to be done in an area that has a high roof, and a space where it doesn't matter if things get a little wet!

Give each of your assistants a film canister, some water, blu-tac and an effervescent tablet. Get them to stand their canisters upright with the lid off. The vitamin C tablet is to be broken into quarters and one piece stuck on the lid using a very small piece of blu-tac. How could the amount of tablet used be measured more accurately?

Now a small amount of water is to be poured into the canister. What could be used to make sure the same amount of water is used each time?

Get them all to fit the lids onto the canisters. A plug lid is best – rather than one that overlaps on the outside. The reaction will not take place until the water touches the tablet.

At a count of 3 get them all to turn their canisters upside down so that the lid is touching the ground, have them stand back and time, using the stop watches, how long it takes for their rockets to lift off.

Now set them a challenge in which there needs to be a 40 second delay – or what ever time you want to set - between the priming of the film canister rocket and lift off! To do this they will have to answer the above questions!

Science in a Nutshell

When an alka-seltza or vitamin C tablet dissolves in water, a chemical change takes place and carbon dioxide gas is formed, this causes the fizz. Alka-Seltzer and Vitamin C tablets contain sodium bicarbonate ($NaHCO_3$), a base, and citric acid ($C_6H_8O_7$), an acid. In the solid tablet the acid and base do not react; the atoms are tightly bound in a crystalline structure and hence can not react with each other. But when placed in water the tablet dissolves, the chemicals are free to move around with the result that the sodium bicarbonate reacts with the citric acid in an acid-base neutralisation reaction.

$$NaHCO_3 \text{ (aq)} + C_6H_8O_7 \text{ (aq)} \rightarrow NaC_6H_7O_7 \text{ (aq)} + H_2O \text{ (l)} + CO_2 \text{ (g)}$$

The tablets can be broken into small pieces, or can even be crushed to become a powder, thus the amount of tablet and particle size can be investigated. The volume of water can also be investigated.

In the closed container, the newly formed CO_2 gas mixes with the air that was already in the canister. The pressure inside the container builds up because more and more gas particles are hitting the sides of the container. This pressure acts in all directions and builds up until the force is large enough to separate the canister from its lid. The gas rushes out, making a whoosh sound.

The film canisters can be propelled up to heights of about 5 m! – The height reached is dependant on how tightly the lid fits the canister. Both this experiment and the upside down 'Volcano bottle rocket' can be used to help illustrate Newton's Third Law of Motion. This law states that for every action there is an equal and opposite reaction; the 'rocket' travels upwards with a force that is equal and opposite to the downward force of the propelling water, gas and lid. The 'rocket' lifts off because it is acted upon by an unbalanced force (Newton's 'First Law'). This unbalanced force which causes the lid to blow off is due to the increased pressure due to the gas formed in the canister. The amount of force is directly proportional to the mass of water and gas expelled from the canister and how fast it accelerates (Newton's 'Second Law'). Phew... lot's of science here!

Fair Test: If you decide to turn this demonstration into an investigation your assistants will need to ensure they understand they can only change one variable at a time. Before you do the experiment get them to come up with their own ideas on what they need to change to make the reaction speed up or slow down. Some examples could be (these are the variables):

1.   amount of tablet (i.e. all, ¾, ½, ¼)

2.   volume of water

3.   temperature of water

4.   container size

5.   particle size (large pieces of tablet or powder samples)

To ensure that they all start the experiment at the same time the tablets are attached to the film canister lids with blu-tac, whilst the water is poured into the canister itself. If powder is used, pour the water in to the canister then place a very thin sheet of tissue paper over the open top; carefully add the powder to the tissue paper making sure none drops into the water. Put on the lid to secure the tissue paper in place.

Your young science investigators can work on their own or in small groups. When everyone in the group is ready to start – have one of them do a count down so that they all turn their film canisters over at the same time.

Fact: Less Gas

If you planning on having baked beans for tea – cut down on the gas produced by all that eat them by adding two teaspoons of baking soda to the cooking pot.

Cabbage Chemistry

Nuts & Bolts

A red cabbage
Saucepan
Water
Sieve
Two large clear plastic beakers
Paper towels!
Two assistants
Two pairs of safety glasses

Safety: Eyewear should be worn by your assistants just in case there is splash back.

Secrets for Success

Chop up the red cabbage, place it in a saucepan and add just enough water to barely cover it. Bring the water to the boil, and simmer for about 15 minutes. Pour the water through a sieve. This purple water is your indicator dye – store in a clear plastic drinks bottle. Label: Red Cabbage Indicator – DO NOT DRINK!

Testing the indicator:

Half fill one of the beakers with vinegar and hand this to assistant 1. Half fill the second beaker with water and add two tablespoons of bicarbonate of soda (baking soda) to it; hand this beaker to assistant 2 getting them to stir in the powder with a spoon or magic wand – which ever is closest to hand! The amount of showmanship is up to you but make sure you highlight the fact that the vinegar is an acid and the baking soda is a base (which when dissolved in water makes an alkaline solution).

Get your audience to guess what will happen when a squirt of the purple coloured cabbage juice is added to each of the beakers. The bicarbonate of soda solution (alkali) will turn the cabbage water a greeny-blue colour. The vinegar (acid) will turn the cabbage water a reddish-pink colour. The cabbage dye behaves as an indicator, a chemical substance which changes colour, depending on the acidity or alkalinity of its environment.

Finally ask which assistant should pour the contents of their beaker into the other assistants' beaker – you could mention here that there is a strong possibility that a reaction will occur and the size of the reaction might depend on which beaker is poured first (? .. just adding to the excitement!) hence the need for safety glasses.

Science in a Nutshell

The purple colour in the red cabbage comes from a class of pigments called anthocyanins; this pigment is also found in the skin of red apples, grapes, plums and is the pigment in leaves which turn red in the autumn.

When the vinegar is mixed with the baking soda solution a chemical reaction takes place. You not only see a colour change; the end solution will be a purple colour – indicating neutralisation has taken place, but LOTS of bubbles of carbon dioxide are formed; these bubbles are as a result of the reaction. Provided that the beakers had enough liquid in them to begin with they will not have a large enough volume to hold the bubbles and they will cascade over the top of the beakers leaving a very wet mess on the floor – hence the paper towels! I actually use a black tray used by builders to mix cement to collect the splashes.

This chemical reaction is described in the Balloon Blow Up experiment.

Note that in the general case of simple acid-base reactions, the term, "salt" refers to the non-water, ionic product. If hydrochloric acid and sodium hydroxide were the reactants, then NaCl (sodium chloride - common salt) would be the non-water product. Just as a reminder our reaction looks like this:

$$CH_3COOH \quad + \quad NaHCO_3 \quad \rightarrow \quad NaC_2H_3O_2 \quad + \quad H_2CO_3$$

Ethanoic acid plus sodium bicarbonate makes sodium ethanoate plus carbonic acid. The $NaC_2H_3O_2$ is the salt called sodium ethanoate. The H_2CO_3 (carbonic acid) then breaks down into water and carbon dioxide:

$$H_2CO_3 \quad \rightarrow \quad H_2O \quad + \quad CO_2$$

This break down is rapid hence the reason why the contents will bubble over the edge of the beaker – most impressive!!

That's pHantastic!

Nuts & Bolts

Red cabbage indicator solution
Filter paper
Scissors
A selection of kitchen foods and chemicals.

Safety: Do not leave young children with any hazardous chemicals. Do not mix any chemicals if you do not know what the reaction between them will be. Some gases formed can be very dangerous. Wear eye protection to prevent splash back into the eyes.

Secrets for Success

Soak the filter paper (or coffee filters) in a concentrated solution of red cabbage juice. After a few hours remove the paper and allow it to dry. Cut the filter into strips and use them to test the pH of a selection of solutions made from, for example, fruit and vegetable juices, toothpaste, alka-seltza tablets, vitamin C tablets, surface and floor cleaners and soap.

Science in a Nutshell

If arranged in order your coloured filter papers should display a spectrum of colours from cherry red (strongly acidic), pink-red (acidic), lilac (slightly acidic), purple (neutral), blue (slightly basic), green (basic) and yellow (strongly basic).

Note that the foods that we eat and drink are mainly acidic and the substances we use for cleaning are basic. Basic substances taste unpleasant but are excellent in helping to remove dirt and grease. Substances that are acidic and basic make the eyes sting, so baby shampoo is made neutral.

Literary Magic

Nuts & Bolts

Baking soda
Water
Cotton buds or small paint brushes
Red cabbage juice
Small spray bottle
White paper

Secrets for Success

Mix equal parts of water and baking soda and, using a cotton bud or paint brush, write a message or draw a diagram onto a sheet of white paper using the baking soda solution as ink. Allow the message to dry.

Pour some of your concentrated red cabbage solution into a small spray bottle (obtainable from a garden centre). Spray this solution over the message. The message will appear in a different colour.

Science in a Nutshell

Acidity is measured on a pH scale which runs from 0 (most acidic) to 14 (most basic or alkaline). A substance which is neither acidic nor basic is called 'neutral' and has a pH of 7. An acid solution contains an excess of hydrogen ions (H^+) i.e. pH is a measure of how acidic a solution is or how many H^+ ions are present. Alkaline solutions have a pH greater then 7 meaning they have less free H^+ ions than that of neutral water.

pH Indicators

pH indicators are dyes which change colour over a range of H^+ ion concentrations. Different dyes will change colour at different pHs. Litmus is a dye obtained from lichen which grows in Northern Europe. It is one of a large number of organic compounds that change colour when a solution they are in changes acidity at a certain point. Litmus is the oldest known pH indicator. It was used to dye wools and other textiles, it is red in an acid solution and blue in an alkaline solution.

Universal indicator is a mixture of pH indicators which gives a range of colour changes depending on the acidity of the solution. The colour change interval is expressed as a pH range.

'Our own kitchen indicator – red cabbage'

Nature uses colour in lots of ways. In animals, colour camouflages – in plants the opposite is more common. Colour in flowers, for example, attracts insects to pollinate the flower. The pigment found in red cabbage is a red colour in acids, a purplish colour in neutral solutions and appears greenish-yellow in strong alkaline solutions. How is this possible?

The colour of the juice changes in response to its H^+ concentration. Acids donate H^+ ions in aqueous (water) solutions and have low pH values. Bases in aqueous solutions (alkalis) however accept H^+ ions and have pH values higher than 7.

As the pH of the red cabbage indicator gets higher there is a loss of H^+ ions. This loss alters the wavelengths of light reflected by the compounds in the indicator, thus creating the colour change with respect to pH.

Colour Scale of red cabbage indicator

Colour	red		violet		purple		blue		green			yellow		
pH	1	2	3	4	5	6	7	8	9	10	11	12	13	14
	Strong acidic		Moderate acid		Weak acid		Neutral	Weak base		Moderate base		Strong Base		

Going Further

How We See the Colour Changes

Light travels in waves called electromagnetic waves. These waves are vibrations of electric and magnetic fields that pass through space. In physics, the visible spectrum has three primary colours: red, green and blue. Chemically, colour is derived from pigments and compounds and the three primary colours here are: red, yellow and blue. Any mixture of these two colours will give a third colour known as a secondary colour.

The diagram below shows what a small part of the whole electromagnetic spectrum light actually forms. We generally refer to the wavelengths of visible light as colour, which we split into seven bands: red (longest wavelength), orange, yellow, green, blue, indigo and violet (shortest wavelength).

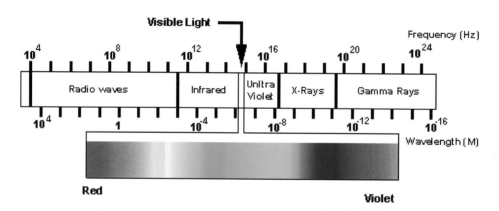

There are two basic ways by which we see these colours. Either an object can directly emit light in the wavelengths of the observed colour, or an object can absorb all other wavelengths, or combinations of light waves, and reflect only those that appear as the observed colour. White light is a combination of all colours whilst black is the absence of colour.

We see colour due to sensors in the retina of the eye called rods and cones. The rods are sensitive to low light and the cones, which require a greater intensity of light, are sensitive to colour. These sensors pass on their activity as a message to the brain via the optic nerve.

As you can see from the red cabbage colour chart a range of colours can be seen. When we add vinegar to the purple coloured cabbage juice we see the colour red. The rods and cones of our eyes pick up this particular wavelength of red because at this concentration the cabbage pigments reflect light rays from the red end of the spectrum and absorb light rays from the blue end.

In theory, any substance can be used as an acid-base indicator if, when there is a change in pH, it undergoes a reversible chemical change. Most materials which we use as indicators are derived from plants. Almost any flower, fruit or plant part that is red, blue or purple contains the class of chemicals compounds called anthocyanins that change colour with pH. Here is a small list of house and garden materials that can be used as acid-base indicators:

Beets change from a red to purplish colour in very basic solutions.

Blackberries and black currents change from a red colour in acid to a dark blue or violet in basic solutions.

Turmeric is a spice that contains a bright yellow pigment called curcumin which, by the way is not an anthocyanin but it does give a nice colour change! It turns from yellow at pH 7.4 to red at pH 8.6. Turmeric is made from ground turmeric tuber and can be found alone or in curry powder.

Onion is an olfactory indicator which means the smell changes rather than the colour. The onion smell isn't detectable in strongly basic solutions. Hence baking soda used in solution can be used to remove the smell of onions from your hands and the fridge.

Chapter 3

Joule be Surprised!

Energy! Such a wonderful word so evocative, full of promises. What we couldn't do if we only had the energy! How many times have you wished to be able to tap into the universal energy surrounding you because all of the worldly and family demands on you have left you sapped; drained; completely exhausted? Oh to be like the humans in the 'Matrix' – How fantastic would it be just to tap into an energy source at will?!

Obviously in this book we are not going to explore ways to increase our own personal energy levels, I will leave that to the fitness gurus, but hopefully the activities you and your wards will engage in will most definitely fill the room with energetic cries of surprise and laughter.

As said in the introduction, we like to categorise information and objects alike, and scientists, being human (believe it or not!) do the same. And this being the case energy can be split into two main categories:

1   Energy that does and

2   Energy that has the potential to do.

The first group is usually labelled movement or kinetic energy and can be sub-divided as follows:

1.1   The motion of waves (light and heat)

1.2   The motion of electrons and ions (electricity and magnetism)

1.3   The motion of atoms, molecules, substances and larger moving objects.

It is the last group which is responsible for sound and the 'occasionally dreaded' equations used to mathematically describe Newton's Laws of Motion!!

The second group, that of potential energy, deals with stored energy. Energy which is found in foods, fuels and batteries in the form of chemical energy, in objects which have been stretched, squashed or wound up, in objects that have been moved from a lower energy level to a higher one for example objects which are magically levitated upwards due to 'The Force' (or hiddens strings!) gain gravitational potential energy and finally, nuclear energy, which is the energy gained due to the position of particles in an atom's nucleus.

To make the science easier we usually split energy into 9 separate systems; however it is very important to stress that these systems are very much interconnected. These energy groups are: Kinetic (movement), Light, Heat, Sound, Electrical, Magnetic, Potential (gravitational and strain), Chemical and Nuclear energy.

Understanding how energy behaves especially on the more microscopic theoretical level is not easy for even those with university science degrees. Present advanced concepts with names such as thermodynamics, entropy, relativity, quantum mechanics, quantum field theory, and string theories are difficult to comprehend. Often complicated formulas and scientific jargon are used. So you will be pleased to know in this book we will only deal with concepts of energy which are useful in simple practical applications in the fields of engineering, physics, chemistry and biology.

Energy powers our vehicles, warms and cools our homes, plays our music, cooks our food and works our toys. All machines, animals, trees and plants change one or more forms of energy into other more useful forms of energy.

An English physicist called James Prescott Joule discovered that heat was a form of energy and as a consequence of his studies in this area the unit for energy, the Joule, was named after him.

When a force moves something or makes it change its shape the energy transfer needed to do that action is called work. Energy allows us to do work and the amount of work done is equal to the amount of energy transferred. No more, no less.

The main law concerning energy is called the Conservation of Energy Law which states that: Energy can not be created or destroyed only changed from one form to another.

Energy makes change; it does things for us. Bakes cakes in the oven, keeps food frozen in the freezer, makes bodies grow and helps keep them warm. It allows us to think, move and play. So let's move on and use some of the pent up energy stored in our bodies to carryout a few more Magical Science experiments!

The Straw Oboe

Nuts & Bolts

Straws
Scissors
Matches
16 sheets of A4 paper
Sound intensity meter
Assistants

Secrets for Success

Using your fingers flatten one end of the straw. Use the scissors to cut the flattened end so it looks like this shape _/, note the angled lengths should be between 1 and 1½ cm in length. Ask your assistants to place the cut end, the reed, into their mouths. Ask them to place the 'reed' onto their lower lip – then with their lips barely touching they have to blow steadily whilst gradually increasing the pressure from their lips until they get a sound. The teeth should not be used!

Note the whole of the cut section must be fully inserted into the mouth. If you are lucky enough one or all of them will get a sound first time round. However, if no sounds are forth coming check that the straw is far enough in the mouth, and or ask them to tighten their lips and gently squeeze the straw with their fingers as they blow – not too much though as air is required to move through the straw!

This demo will produce lots of laughter from everyone – the flattened _/ shaped tip acts like the reed found in most wind instruments. Blowing on the reed causes the straw to vibrate. The vibrations are felt by the lips of the blower and they usually stop blowing first off because the experience is very weird. The buzzing sound is a bit like a musical duck call!

Once you have an assistant that can give you a long note, get them to hold the straw close to their mouth. With the scissors try cutting the end off of the straw, notice what happens to the change in the sound of the note. Keep cutting until the straw is a third of its original length – let your assistant know what you are going to do as they will get fearful for their fingers!!

Repeat the process again but this time with the aid of 6 matches melt 6 holes in the straw about 1 ½ cm apart, starting 1 ½ cm up from the un-cut end. You can now play your straw oboe!

Let's try Twinkle Twinkle Little Star: the number 1 represents the top hole covered, 2 the top two, etc., until 6 which means all holes are covered.

6 6 2 2 1 1 2,   3 3 4 4 5 5 6,   2 2 3 3 4 4 5,

2 2 3 3 4 4 5,   6 6 2 2 1 1 2,   3 3 4 4 5 5 6

Science in a Nutshell

This is a lovely little demonstration which can be used to aid understanding in topics such as materials, sound, waves, energy transfer and the relationships between wavelength, frequency and the speed of the wave ($v = f\lambda$), and frequency and pitch.

When the flattened end of the straw is cut into a V shape and placed into the mouth, the '_/' shaped reed vibrates when you blow gently, this is because when you blow a pulse of compressed air flows down the straw. The pulse travels down the straw at mach 1, the speed of sound, and bounces off the distant open end. At this point the compressed air changes into a low pressure expansion (low pressure because, due to the expansion of the air, there are less air molecules per unit volume, in other words the density of the air has decreased). When the expanded air reaches the two flattened edges of the straw they are forced closed they then bounce open again to admit more air. Thus the sound bounces back and forth inside the straw and the flattened edges, our reed, opens and closes (vibrates) creating the duck like sound.

How does sound bounce off an open end of a tube?

This can be demonstrated firstly with the aid of a bottle brush and a tube. Push the brush inside the tube (could be made from a piece of rolled up card) – when the brush exits the tube its' bristles spring outward this movement demonstrates what the compressed air in a sound wave does. But air is more elastic than bristles. A large tube (PVC) and a lead weight which has one end tied to a long spring of rubber bands can be used to illustrate the compression overshoot. Without letting go of the rubber band (!) drop the weight down the tube. When the lead weight comes out of the end of the tube, it stretches the rubber bands and hence overshoots its equilibrium position before bouncing back up the tube, just as air does. The air compression overshoots just like the lead weight and pulls a partial vacuum at the end of the tube which goes back down the tube as an expansion; that's how sound bounces back and forth inside a tube.

Now it's your turn to show off! Once you are able to produce a clear sound, cut off successive pieces of the open end of the straw. You will hear the pitch of the note increase with each cut. This increase in pitch is due to the change in wavelength (the symbol used by scientists for wavelength is λ) of the standing wave set up in the straw. With each cut the straw is made shorter which means the wavelength of the sound wave becomes smaller. Mathematically the speed of sound, v, the frequency of the note, f, and the wavelength of the wave, λ are related to each other by the following formula: $v = f\lambda$

In this experiment the speed of sound is a constant, it can not change, so by cutting the straw we decrease the wavelength, λ which means, for the formula to hold the frequency of the note i.e. the number of times the reed opens and closes per second, must increase. We perceive this change as an increase in pitch of the note.

Has the formula confused you? Try putting numbers into the equation, for example 36 = 2 x 18. If the number 18 is decreased to 6 and the number 36 is to remain unchanged, the number 2 must change [2 x 6 = 12 not 36!] It must be increased to the number 6 [6 x 6 = 36].

The loudness of the note can be increased if you blow harder or by attaching a paper cone (use poser paper if possible) to the end of the straw. Try a cone firstly made from one sheet of A4 paper then 4 and lastly 16 sheets.

In schools the changes in loudness can be detected using a sound intensity metre and loudness and pitch (frequency) can be investigated with the help of an oscilloscope and a microphone. Students can see the amplitude of the sound wave on the screen of the oscilloscope increase in height as the loudness of the produced note is increased. When the length of the straw is changed the increase in pitch is registered on the oscilloscope by the distance between successive wave crests moving closer together.

By adjusting the tension in your lips and the pressure with which you blow different harmonics can also be played. The straw is an open ended tube, so when you blow and hear a note this means a standing wave has been set up in the straw. The fundamental wave will have a wavelength = 2L where L is the length of the straw. The wavelength of the second harmonic is equal to L and so on. Illustrations of displacement and pressure patterns in open pipes can be found in many A-Level Physics text books: happy reading!

Let's Investigate

Give your assistants a set of straws and ask them to answer the following questions:

How can you increase and decrease the pitch of the straw oboe? (*Change the length by cutting or by joining successive straws together*)

Why do you have a different pitch when the length varies? (*Link to standing wave diagrams in text book and they should see that the wavelength is different for different straw lengths – since v remains constant, f must change*)

Where is the least air pressure in the straw oboe? (*At the wave antinodes where the air molecules are spread far apart the pressure is least – conversely, the air molecules are all bundled together where the wave nodes are.*)

Going Further

Cut a 50 cm length of sticky tape and place the centre of 25 straws onto the tape at 2 cm intervals. Stick a second 50 cm length of tape over the top of the straws to cover the sticky part of the lower tape. Attach a paper clip to each end of every straw. Hang the straw strip from a desk, pull it taunt, and give one of the straws at the top or bottom a tap to start a transverse wave. Increasing tension will increase the speed of the wave, and increasing the density (by adding more paperclips) will decrease the speed of the wave. Although sound is a longitudinal wave, best represented by compression waves in slinky springs, the straw waves can be used to illustrate many of the properties of sound waves.

Links to Chemistry: What do reeds have to do with polymers?

Many of the materials used to make reeds are polymers. Polymers are flexible and able to absorb moisture, which makes them great to use for reeds. Most reeds are made from cane, which is a natural polymer, but these days there are many synthetic polymers being used to make reeds as well. Natural reed materials, such as cane, can be unpredictable and unstable especially when they are too wet, too dry, or too weak. Scientists have made new reed materials out of polymer composites which can very closely imitate cane as to how they react to air and what sounds they produce, they are longer-lasting and easier to keep in good playing condition.

The Big Glass Band

Nuts & Bolts

Six concave wine glasses, same size
Measuring jug - small
Water
Food colouring or paint
Metal spoon
Egg cup filled with vinegar

Secrets for Success

Pour water into the glasses so that you have six different levels. Add food colouring or paint to each glass so that you can easily see the water levels – try to get red, orange, yellow, green, blue and purple.

Tap each glass with the spoon to create different sounds. The glass with the most water will produce a note much lower than the one with the least water in it; the pitch of the note increases as the water volume decreases.

Does it make any difference if you hit the glass with a metal, plastic or wooden spoon?

Does the surface the glasses stand on affect the sounds heard?

Can you adjust the water levels in the glasses so that you can play a tune that you know?

Colour	Volume of water (ml)
R Red	310
O Orange	280
Y Yellow	200
G Green	150
B Blue	80
P Purple	30

Using your spoon hit the glasses in the following sequence; can you guess the tune? Vary the volumes until the notes are clear.

R R B B P P B

G G Y Y O O R

B B G G Y Y O

B B G G Y Y O

R R B B P P B

G G Y Y O O R

Now wash your hands with soap to clean away any dirt or oils. Choose one of the glasses, hold it steady with one hand, dip a finger in some vinegar and then keeping contact with the glass at all times, move your finger at a constant speed around the rim of the glass, vary the speed until you hear a loud, clear note.

Repeat with the other glasses. What happens to the pitch of the note as the volumes of water in the glass increases or decreases?

Find a way to secure the glasses to the table; blu-tac, sellotape or build a holder. Can you play the same tune (which hopefully you noticed was Twinkle, Twinkle Little Star) or the tune of your choice by rubbing the rim of the glasses with your fingers?

Science in a Nutshell

The vinegar is used to ensure there is no grease on your finger tips. Best effects are achieved if a steady pressure is applied between a clean finger and the glass, this is because it allows you to get the stick-slip motion required to produce sound.

Sound is usually considered to be a vibration having a range of frequencies that can be detected by the human ear. This is between 50 Hz (50 vibrations a second) and 20 000 Hz (a staggering 20 000 vibrations a second).

Different materials vibrate differently. For us to hear a sound produced by vibrations the sound waves generated need to travel to our ear drums. The speed at which sound travels is also dependant on the material it is travelling through. The speed depends on how closely packed the particles in the medium the sound is travelling through are.

In gases the particles are far apart, the speed of sound in air is between 300 and 350 m/s. Sound travels faster in damp or polluted air. Why do you think this is the case?

Dominoes (or Jenga pieces) can be used to illustrate how sound is transmitted to our ears without the transfer of mass. Arrange as many dominoes as possible in a row, knock the first down and ask your audience to describe the disturbance as it passes through the dominoes. Does it travel at a constant speed? (It should do!) If you have enough domino shaped pieces you could organise your assistants to work in teams. Get them to investigate what happens to the speed if the dominoes are closer together (speeds up) or further apart (slows down). Then link this to the increase in the speed of sound as the density of the material through which it travels increases.

In water, sound travels at speeds between 1200 m/s and 1500 m/s. It travels faster in sea water than in fresh water. Why is this? Sea water has lots of salts dissolved in it which means it has a higher density than fresh water.

The speed of sound in steel is around 5000 m/s, approximately 15 times faster than the speed of sound in air.

Fact

Mozart wrote music for an instrument called a glass harmonica. It consisted of some glasses on a tray, each with different amounts of water in them!! He played them by rubbing a finger around the rims of the glass – so how did your tune compare with those written by Mozart?? Who by the way was the composer of Twinkle Twinkle Little Star.

Brring, Brring, Brring – it's a Telephone made from String!

Nuts & Bolts

2 plastic cups or empty yoghurt pots
Some string
2 paper clips
Sharpened pencil or object for poking holes.
An assistant

Secrets for Success

Loosely hold the string and gently move your thumb along its length. What do you hear? Not much. Now hold the string taut and repeat. Is the sound slightly louder?

With the pencil, carefully poke a small hole in the bottom of each container. Tie the paper clip to one end of the string. Thread the other end through the hole moving from the inside of the cup outwards. The paperclip will keep the string from going all the way through the hole. Hold the container in one hand and gently drag the thumb nail of your other hand along the string moving from the top downwards. What do you hear this time? You will hear a much louder sound.

Thread the string through the hole in the second container, this time from the outside to the inside. Tie the second paper clip to the end of the string. The paperclips should both be on the inside of the containers.

Hand one of the containers to your assistant. Move apart so that the string between you is taut. Ask your assistant to put their container to one of their ears while you talk into your container. Make sure that the distance between you both is sufficient for your assistant not to be able to hear what you say without the aid of the 'telephone'!

What is the farthest distance you can get the telephones to work using string?

Does using thin metal wire affect this distance?

Does increasing the size of the cup affect the volume of the sounds you hear?

Science in a Nutshell

When you drag your nail across the string, the rough surface causes your nail and string to vibrate. The container, with its large surface area, amplifies this sound. Larger containers will give you louder sounds because more air is available to vibrate hence more energy will be transferred.

Going Further

Many toy shops or trinket shops such as 'Past Times' sell music boxes. Little mechanical devices with handles which when turned cause a central metal tube with spokes on to turn. The spokes then hit against a selection of metal prongs which then vibrate up and down emitting a sound. Hold the music box in the air. Turn the handle. Can everyone in the room hear the tune being played? Now place the box onto the base of a plastic beaker. Turn the handle. Wow! The sound is significantly louder. Try placing the music box on to a range of surfaces – which one gives you the loudest sound.

The body of a guitar is similar to the cup – it is a sound box. It is shaped so as to not only amplify the sound but to enhance the sounds heard. You can make your own guitar by putting elastic bands around a shoe box. Thick and loose bands will give you low notes, thin and taut bands will give you higher notes. The tension of the strings on a guitar is varied using the tuning keys which are found at the top of the finger board.

Good Vibrations

Nuts & Bolts

Metal spoon
Metal coat hanger
Wooden spoon
1 ½ metres of string

Secrets for Success

Tie the metal spoon to the centre of the string. Wrap one end of the string about 5 times around the index finger of your left hand and the other around the index finger of your right hand. Put the tips of both fingers into your ears. At this stage you may feel a bit silly if others are in the room watching you but the experience will be worth it and all will want to try!

Now bend forward allowing the spoon to strike against different surfaces – you will hear loud, alarming, chiming sounds! Repeat using other objects such as a coat hanger or wooden spoon.

Science in a Nutshell

The object attached to the string will vibrate when it is hit. The sound waves are transmitted to the ear drums via the string and your fingers. This activity demonstrates that solid materials are much more effective at transmitting sound energy than air, which is a gas.

Fact:

Loud concerts can cause temporary shifts in the threshold of hearing in the mid frequency region. There is no firmly established correlation between temporary threshold shifts and permanent threshold shifts. However, it is prudent to point out that if permanent hearing damage followed the pattern of the temporary threshold shifts, it would be the worst kind of damage because it would diminish hearing acuity in the frequency range that is most important for the understanding of human speech. This is why you will sometimes find older people who can hear whispers and the TV when it is on maximum sound levels but find listening to conversations in a room difficult with the help of a hearing aid.

Sparkling Lines

Nuts & Bolts

Hand drill
Long sparklers
Lighter

Safety: This is an activity to be done outside because cascading sparks present a fire and burn hazard. Use them safely and have fun! – Perfect demonstration for Hallowe'en parties.

Secrets for Success

Secure the sparkler in the drill; bend it close to the drill bit through an angle of 90 degrees. Light the sparkler and start turning the drill handle slowly then gradually speed up. Ask your audience to focus on the 'sparkles' leaving the drill. Get them to describe the movement of the sparkles when the handle is first turned slowly and then quickly. Look at the initial path the sparkles take once they leave the surface of the sparkler. They leave at right angles to their original circular path and for a period move in straight lines, though eventually gravity (our constant external force) wins and the sparkles fall to earth.

Science in a Nutshell

The sparkles fall to earth because of the gravitational force of attraction between them and the earth. If the sparks were in outer space, away from the influence of any planet or stars' gravitational field, they would in fact keep moving in a straight line – if Newton's laws were to be held true.

However Albert Einstein threw a spanner in Newton's works about 90 years ago; he stated that space is actually curved! This would make straight line motion very difficult! But in practical experiments on earth, Newton's Laws of Motion hold up very well.... But who is to say that this will not change – so many ideas which we thought and taught to be right have changed. It wasn't that long ago when it was thought that the world was flat and any anyone saying otherwise was tortured!!!

What is a Sparkler?

A sparkler consists of a chemical mixture that is molded onto a rigid stick or wire. These chemicals often are mixed with water or alcohol along with a binder to form a slurry that can be coated on a wire (by dipping) or poured into a tube. Once the mixture dries, you have a sparkler. Sparklers are designed to burn over a period of 1 or 2 minutes. They produce a brilliant shower of sparkles.

Aluminium, iron, steel, zinc, titanium or magnesium dust or flakes may be used to create the bright, shimmering sparks. The metal flakes heat up until they are incandescent and shine brightly or, at a high enough temperature, they actually burn. A variety of chemicals can be added to create colours. A fuel such as charcoal or sulphur and an oxidiser are proportioned, along with the other chemicals, so that the sparkler burns slowly rather than exploding like a firecracker. Once one end of the sparkler is ignited, it burns progressively to the other end.

Now that we know what they are made of let's describe the chemical reaction in a little more detail. Hopefully by now you will be able to understand some of the scientific shorthand used to represent a chemical though I will use both long hand and short hand notes to describe the reaction!

The oxidisers are usually nitrates, chlorates, or perchlorates; they produce oxygen during the burning process to help keep the mixture burning. Nitrates are made up of a metal ion, say for example a potassium ion (K^+) and a nitrate ion (NO_3^-). Nitrates give up 1/3 of their oxygen to yield nitrites (NO_2^-) and oxygen (O_2). The resulting short hand equation for the potassium nitrate reaction looks like this:

2 KNO$_3$ (solid) → 2 KNO$_2$ (s) +O$_2$ (gas)

Two potassium nitrate molecules during combustion give two potassium nitrite molecules plus one oxygen molecule. Both the nitrate and nitrite are solids and the oxygen is a gas.

Chlorates are made up of a metal ion and the chlorate ion. Chlorates give up all of their oxygen, causing a more spectacular reaction. However, this also means they are explosive. An example of potassium chlorate yielding its oxygen would look like this:

2 KClO$_3$ (s) → 2 KCl (s) + 3 O$_2$ (g)

2 potassium chlorate molecules → 2 Potassium chloride plus 3 oxygen molecules.

Perchlorates have more oxygen in them, but are less likely to explode as a result of impact, caused say by them being dropped, than are chlorates. Potassium perchlorate yields its oxygen in this reaction:

KClO$_4$ (s) → KC l(s) + 2 O$_2$ (g)

1 Potassium perchlorate molecule → 1 potassium chloride and two oxygen molecules.

The fuel used to burn the oxygen produced by the oxidisers is also known as the reducing agent. This combustion produces a hot gas. Examples of reducing agents are sulphur and charcoal, which react with the oxygen to form sulphur dioxide (SO$_2$) and carbon dioxide (CO$_2$), respectively.

Two reducing agents may be combined to speed up or slow down the reaction. Also, the sizes of the metal particles also affect the speed of the reaction. Finer metal powders react more quickly than coarse powders or flakes. Other substances, such as cornmeal, may also be added to regulate the reaction.

Binders hold the mixture together. For a sparkler, common binders are dextrin (a sugar) dampened by water, or a shellac compound dampened by alcohol. The binder can serve as a reducing agent and as a reaction moderator.

Further still ...

So how is motion in a circular path then explained??

Motion along a curve or through a circle is always caused by a force that pushes the moving object in an inward direction – we call any force that does this a centripetal force. These forces are used in washing machines, on many rides found in theme parks, and are most useful for us when in a car or on a bike wishing to turn left or right!

For any vehicle to turn a corner there must be a force acting towards the centre of the circular path taken. This force is provided by the frictional contact forces between the wheels and the ground. No frictional forces, no turning corners! This is observed frequently during spells of icy cold weather! In this situation Newton's First Law prevails; you and your mode of transport will continue moving in a straight line unless an external forces acts upon you – no friction, no external force. Hence it is always wise to travel very slowly and carefully in icy conditions – if you end up colliding with a wall with a small speed your momentum change (when you come to a stop!) will be small hence the forces acting on you will also be small.

This demonstration is a nice way of visualising Newton's Second Law; the more force applied the faster the object moves, and his First Law which states motion will be in a straight line unless an external force acts upon it. The metal pieces stay on the un-burnt section of the sparkler because of strong binding and frictional forces. As soon as combustion starts these forces are weakened and the metal pieces are free to move, and as you can clearly see their initial path is in a straight line!

The Measure of 'c'
by M Mallow

A strange title but I thought 'Measuring the Speed of Light Using Marshmallows' was too long!

Nuts & Bolts

Microwave oven
Bag of large marshmallows
Ruler
Calculator
Microwaveable plate

Secrets for Success

Remove the turntable from inside the microwave. Completely cover the plate with a layer one marshmallow thick. Cook the marshmallows in a low heat until they begin to melt in 4 or 5 different places. Remove the plate from the oven and with a ruler measure and note down the distances between the melted spots. You should notice one distance is repeated over and over again. Turn the microwave round and look for a small sign on the back of the machine that tells you the frequency of the microwaves used - most commercial ones operate at 2450 MHz (2 450 000 000 Hertz).

Science in a Nutshell

Microwaves are part of the electromagnet spectrum (gamma rays, X-rays, ultra violet, visible light, infra red, microwaves, radio waves) they have a longer wavelength than light but a much shorter one than commercial radio waves. All electromagnetic waves travel at the same speed which we call the speed of light = 3×10^8 m/s (300 million m/s) the symbol c is used to represent this special speed.

Microwave ovens cook unevenly, which is why the turntable is usually required, to ensure even cooking. We however want to use this phenomenon to our advantage so the turntable is removed. The microwaves are generated by a device at the back of the machine and then directed into the oven cavity. Here they are reflected off of the inside walls and then interfere with each other. Standing waves are formed inside the oven chamber due to this interference and the pattern creates an array of hotspots throughout the oven's volume. The hotspots are in a complex 3D pattern and occur at intervals of every half wavelength.

The lengths that you measured are equal to half wavelength quantities and should be between 0.05 m and 0.07 m. This will give you an approximate value for the wavelength of microwaves used by your oven between 0.10 m and 0.14 m (0.122 m would be magical!!).

To calculate the speed at which they move we must use the following formula:

Speed (in m/s) = Frequency (in Hertz) x Wavelength (in metres)

Thus with the above numbers:

Speed = 2 450 000 000 x 0.122 = 300 000 000 m/s

It will be quite amazing if you got this exact value!! But it should be pretty close. If you want to calculate your percentage error, use the following formula:

[(300 000 000 – experimental value) / 300 000 000] x 100 = % error

See Literary Magic for more information on the electromagnetic spectrum, light and colour.

Up Periscope!

Nuts & Bolts

Two 1 litre milk cartons
Two small flat pocket mirrors
Sharp knife or scalpel
Ruler
Pencil or pen
Masking tape

Safety: Care must be taken with sharp knives. If activity is to be done with young children do the cutting for them.

Secrets for Success

Use the knife to cut around the top of each milk carton, removing the peaked "roof." Cut a hole at the bottom of the front of one milk carton. Leave at least 1 cm of carton on each side of the hole – this is the section that you will look through. Put the carton on its side and turn it so the hole you have just cut is facing to your right. Using the ruler measure the width of the carton. On the side that's facing up, measure this width distance up the left edge of the carton, and use the pencil to make a mark there. Now, use your ruler to draw a diagonal line from the bottom right corner to the mark you made, this line should be at an angle of 45 degrees to the base of the carton.

Starting at the bottom right corner of the carton, cut on that line, don't cut all the way to the left edge of the carton-just make the cut as long as one side of your mirror. If your mirror is thick, widen the cut to fit.

Slide the mirror through the slot so the reflecting side faces the hole in the front of the carton. Tape the mirror loosely in place.

Hold the carton up to your eye and look through the hole that you cut. You should see the ceiling, or what ever is above you, through the top of the carton. If what you see looks tilted, adjust the mirror and tape it again.

Repeat the above with the second milk carton. Stand one carton up on a table, with the hole facing you. Place the other carton upside-down, with the mirror on the top and the hole facing away from you. Use your hand to pinch the open end of the upside-down carton just enough for it to slide into the other carton. Tape the two cartons together.

Now you have a periscope! If you look through the bottom hole, you can see over fences that are taller than you. If you look through the top hole, you can see under tables. If you hold it sideways, you can see around corners.

Science in a Nutshell

Periscope comes from two Greek words, peri, meaning "around," and scopus, "to look." A periscope lets you look around walls, corners, or other obstacles. Sub-marines have periscopes so the sailors inside can see what's on the surface of the water, even if the ship itself is below the waves.

When you are making a periscope, it's important to make sure that your mirror is positioned at a 45-degree angle.

How does your periscope work?

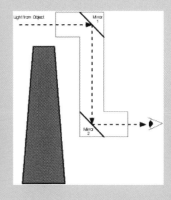

Light always reflects away from a flat mirror at the same angle that it hits the mirror. In your periscope, light hits the top mirror at a 45° angle and reflects away at the same angle, which means the light bounces down to the bottom mirror through an angle of 90°. That reflected light hits the second mirror at a 45° angle and is reflected away at the same angle, 45°, straight into your eye.

Chapter 4

Properties of Materials

Rather than discuss the various properties of a different selection of materials and objects I thought I would share with you some thoughts and discussions on the combination of just three examples of matter, these being: bread, butter and a cat. We all know that if we drop a buttered piece of bread, it drops butter side down and if we drop a cat it always lands on its feet, So what would happen if we strapped a piece of buttered bread to the back of a cat and dropped it out of a first floor window?

To begin with we need to understand the source of the forces that are acting. The force acting on the bread is not the butter, as some may think. Without the bread, butter wouldn't land bread side up, and therefore the force could not possibly be in the butter. We know the force is not the bread because it has been experimentally proven that bread does not land any particular side down without butter. It is obviously the force invoked due to the fusing of bread and butter particles together. This fusion causes energy to be released in the form of shifting gravity and anti-gravity energy to opposite sides of the bread/butter continuum. The gravity energy naturally shifts to the butter since it is denser then the bread, while the anti-gravity energy shifts to the bread side.

The energy in a cat for landing on its feet comes from the feet themselves. This has also been proven experimentally. Cats without feet have a near zero success rate of landing on their feet. We will call this energy 'cat foot energy'.

Considering the equal but opposing bread/butter and cat foot forces one would expect the cat to spin violently about its axis. The engineers and physicists working on this problem have mentioned in numerous publications that due to the strength of these forces they can not be ignored because a regular cat is not structurally stable enough to withstand the torque the spinning causes. They are in fact so strong they can cause the cat's limbs to give way, the legs have been known to be wrenched around until the feet are on the same side of the cat as the butter. And thus the cat can then land on its feet, butter side down.

Even if you are unable to do this experiment yourself you should be able to deduce the obvious result. The laws of butterology demand that the butter must hit the ground, and the equally strict laws of feline aerodynamics demand that the cat can not smash it's furry back. If we are able to overcome the problem of keeping the structure of the cat stable and if the combined construction were to land, nature would have no way to resolve this paradox. Therefore it simply does not fall.

That's right you, your thoughts are correct! You have discovered the secret of antigravity! A buttered cat will, when released, quickly move to a height where the forces of cat-twisting and butter repulsion are in equilibrium. This equilibrium point can be modified by scraping off some of the butter, providing lift, or removing some of the cat's limbs, allowing descent.

As a result of these investigations material physicists and chemists are now working with the Institute for Alternative Energy Research. They are researching the possibility of using structurally reinforced cats for levitation systems, but so far the cost is too high to be practical. Several attempts at producing economically viable systems were made by separating the feet so that the instability of the cat would not be a factor. At first there was difficulty because there was no cat to tie the bread to. Later it was discovered that when not attached to a cat the feet lost their cat foot force over time. It is hypothesized that the feet need to be living to exert the cat foot force, and so far no practical method has been found for keeping the feet alive other than attached to a cat.

Attempts are also being made to breed flat cats with no legs (only feet).

There are many other problems related with this method of levitation as you may well imagine, but they are beyond the scope of this discussion.

Due to an unfortunate misunderstanding (flora was used instead of butter!) several post within this research team have become vacant. For more details visit www.xs4all.nl

The Fireproof Balloon

Nuts & Bolts

Drinks bottle with pop-up lid
Water
Dark coloured balloon
Fire lighter (one with a long handle)
Apprentice

 Safety: Care must be taken with naked flames.

Secrets for Success

Half fill the bottle with water. Push the pop-up lid down. Partially blow up the balloon and very carefully, without losing any air, place the neck of the balloon over the pop-up lid. Turn the bottle up-side down. Open the lid and squirt water into the balloon – about half a cup full. Push the lid in again, remove the balloon carefully (!) then continue blowing up the balloon until it is ¾ of its maximum size. Tie it shut. This is all done behind the scenes.

Secure the balloon with the aid of an elastic band to the end of a stick and then hold the balloon over the head of your apprentice. Give it a shake. Ask 'What do I have inside this balloon?' then 'What can I do to the balloon to allow this magical liquid to escape?' You will receive lots of shouts relating to burst it with a pin... but pull out a cooks lighter. Light it. They will squeal with delight knowing that you are going to burst it with fire. Your volunteer at this stage will look slightly perplexed!

It is at this point you could introduce an umbrella. Your volunteer will be pleased - your audience will groan. It is up to you whether you keep the umbrella or discard it!!

Now click your lighter, slowly move the flame up under the balloon until the flame is touching the skin of the balloon. Your audience will have gone silent. They will be confused! What should the heat energy do to the balloon?

Do the above with as much flare as you can muster, but make sure the flame only touches the under section of the balloon in contact with the water. Allow the flame to touch the skin of the balloon for several seconds. You will start to see soot forming on the underside of the balloon but magically the balloon doesn't burst.

Now ask your very brave apprentice to step to the side and hold the stick attached to the balloon. Repeat the demonstration, but this time slowly work your way up from the bottom of the balloon, past the water level. At this point the balloon will burst and water goes everywhere!

Science in a Nutshell

Usually the rubber of an inflated balloon is strong enough to resist the increased air pressure inside the balloon – provided you don't put too much air inside! However, the heat from the flame should weaken the rubber to a point where it is too weak to resist the air pressure and the balloon bursts. The water inside the balloon prevents this from happening because it absorbs most of the heat from the flame – it conducts the heat away from the rubber and therefore the rubber doesn't weaken – unless of course you accidentally go beyond the water margin which would lead to your apprentice becoming very wet!! Oops!

Water is a very good absorber of heat energy; it takes 10 times as much heat energy to raise the temperature of 1 g of water by 1 °C than it does to raise the temperature of 1 g of iron by 1 °C. This is why it takes such a long time for a kettle filled with water to come to the boil. The bonus of this phenomenon is that water releases a lot of heat into the surrounding atmosphere when it cools. This is why areas near oceans and large lakes do not get as cold in the winter as areas at the same latitude further in land.

As with a lot of the demonstrations in this book this experiment can be used to illustrate energy changes, Newton's Laws of Motion as well as the properties of materials.

When you start blowing into the balloon the balloon starts to get bigger, the skin starts to stretch to facilitate the added air. Potential energy is stored in this stretched rubber. We can use this energy to demonstrate the Law of Conservation of Energy: Energy can not be created or destroyed just changed from one form to another.

If you just keep blowing into the balloon there will come a point when the pressure inside the balloon is too high for the rubber material to support. The skin weakens rapidly and the balloon bursts. The stored potential energy is converted mainly into sound energy but also into kinetic energy of the moving pieces of the broken balloon and a little bit of heat energy.

If you are scared of loud noises you could just blow up the balloon and release it, some balloons are specially made for this activity and make lots of noise as they fly around the room gradually getting smaller and smaller. As they move they sometimes gain height, which means they gain gravitational potential energy, this is converted back into motion energy as they fall again. The balloon used in this fashion is a bit like a miniature rocket.

When you let go of the neck of the balloon why does the balloon move forwards? Remember Newton stated that for a body to move there needs to be resultant force acting on that body in the direction of the motion. What pushes the balloon forwards? It is not, as some text books say, the air leaving the neck of the balloon. That gas is moving in the opposite direction and pushes on particles (not contained by the balloon) causing them to move in the opposite direction to the direction the balloon is moving in!! So how can it push on the balloon? The answer is it can't!! So what does?

The demonstration I use to illustrate this paradox is 'borrowed' from a physics lecturer at Queen's University Belfast - so my thanks are given. You will need a cardboard box, some Velcro, scissors and a long stick. The box needs to be open at the top and with the help of the scissors one of the narrower sides needs to be cut down the middle from the top to the bottom and then the bottom edges also cut, leaving you with two flaps.

With a piece of cardboard taken from the top of the box make a rectangular strip 10 cm in width and the same height as the box. Add Velcro hooks to the insides of the flaps and Velcro loops to the strip of cardboard in a fashion that will allow you to close the flaps securely.

With them closed, place the box on a bench, or the floor, and with the aid of the stick quickly hit the inside walls. Note if one wall is hit more frequently than the others the box will move in that direction. You want to hit the walls so that that is no overall movement of the box – this represents the molecules of air hitting the insides of the tied balloon. The pressure due to the collisions of the many billions of molecules in the balloon acts in all directions equally. However, when you open your flaps by removing the velcroed strip, the pressure now does not act equally. This is the same as untying the neck of the balloon. When you now hit inside the box the forward hits are now not balanced by the backward hits which means you have a resultant force acting forwards so the box will move forwards. The balloon moves forwards due to the resultant force from the air molecules still trapped inside of it!

Disappearing Water Trick

Before we do this trick we have to collect the off white crystals found in babies nappies! Using scissors cut around the outside of a nappy and place the middle section into a large Ziploc™ bag. Carefully separate the layers then seal the bag. Shake for about 1 minute; you should see the crystals start to collect in the corners of the bag. Slowly open the bag; bin the large pieces of material and the fluffy fibres. Store the granules in an airtight container. Wash your hands.

The chemical name for these crystals is sodium polyacrylate. If you have a garden centre close to you can buy small packets of these crystals. Here they are usually called water loc or super gel. These crystals can magically (and rapidly) absorb 500 -1000 times their mass in water! However they release the trapped water slowly which is why they are mixed with the soil used in outdoor hanging baskets.

Nuts & Bolts

3 polystyrene cups (or mugs which are white inside)
1 teaspoon of sodium polyacrylate crystals
Jug of water

Safety: If inhaled sodium polyacrylate can irritate the nasal membrane. Avoid eye contact by wearing safety glasses; if it gets into eyes they will become dry and irritated. Be sure to always wash your hands after handling this chemical. Dispose of the used crystals in a bin, not down a sink.

Secrets for Success

Place the granules in one of the cups; pour in a small amount of water to 'set' the gel. Do this away from your audiences' eyes; but just a few minutes before you do the trick. Turn the 3 cups upside down on the table. With your audience in front of you turn the cups the right way up very quickly showing them the insides of the cups – the gel doesn't show up against the white background. Now pour in about 10-20 ml of water. Move the cups about on the table and ask your audience to guess which cup the water is in. Let them know that which ever cup they choose you will turn up-side down over your head... but if they choose the wrong cup you should be allowed to throw the water over one of their heads!!! Fairs fair.

Your audience will obviously choose the right cup; will then be amazed to find that it is empty and then will very quickly get agitated as they realise that one of them may get a soaking!! After all of the 'pretending to get wet' drama and the realisation that all of the cups are empty – some may shout out that you have a sponge or tissue in the cup – these choices would in-fact absorb the water but because they are not sticky will fall out of the cups if turned up-side down. Put your fingers into the cup with the gel; scoop some up and let them see it. Make sure you wash your hands afterwards – or wear disposable gloves.

Science in a Nutshell

Sodium polyacrylate is a polymer; 'poly' means many and 'mer' is a unit of something, in chemistry this is usually a molecule – so polymer literally means 'long chain of molecules'. The sodium polyacrylate crystals soak up water using the process of osmosis (water molecules pass through a barrier from one side to the other). The polymer chains have an elastic quality; but there is a limit to the amount they can stretch and hence a limit to the amount of water they can hold.

One of the greatest uses of sodium polyacrylate is in making nappies super-absorbent. Table salt (NaCl) destroys the gel and releases the water. How?? No one knows – so if you want to make a name for yourself why not study the chemistry of this compound and become the next science superstar!

Polymers, long chains of chemicals, are found everywhere in nature; proteins are polymers. These are found for example in hair, feathers and cartilage. Cellulose is also a polymer which is found in wood, leaves and other plant parts. Polymers can also be made by man; plastic and nylon are good examples of man made polymers.

Going Further

Try the following experiments: How much water will a Pampers' nappy hold? How do other brands of nappies compare? How much water does a teaspoon of sodium polyacrylate hold?

Insta-snow is a different form of sodium polyacrylate. It comes as a white powder and is usually found in toy shops in and around Christmas time. This powder magically transforms from the white powder to a super fluffy substance that looks and feels like snow in seconds! It is so realistic it is now used on movie sets and in indoor snow-boarding parks. You can also buy this powder from the following website: stevespanglerscience. com

Ice-cream in a Bag

Nuts & Bolts

Medium Ziploc™ bag
Small Ziploc™ bag
¼ cup of milk
¼ cup of whipping cream
3 tablespoons of sugar
¼ teaspoon of vanilla flavouring
¾ cup of sea salt
enough ice cubes to ½ fill the medium bag
spoon
thermal gloves
Apprentice
Salsa music – optional!

Safety: The bag is not to be held by unprotected hands as the contents are cold enough to cause tissue damage! If you don't have any thermal gloves you can have your apprentice just hold the top of the larger outer bag or you can replace this bag within a catering size coffee can which can be rolled back and forth on the floor. Gloves or tea towels are needed to protect the hands from the cold – more ice and salt will be required.

Secrets for Success

With the help of your apprentice put the following ingredients into the small Ziploc™ bag: milk, cream, sugar and flavouring. Securely seal the bag ensuring that as much air as possible has been removed from the bag at the same time. Shake the bag to mix the ingredients.

Half fill the larger bag with crushed ice cubes; put the sea salt and the smaller bag on top of the ice. Close this outer bag securely. Ask your apprentice to put on the thermal gloves; they are to shake the bag GENTLY from side to side for 3 minutes (or until the milk mixture in the small bag has solidified). This can be done to the rhythm of salsa music if you wish!

Remove the small bag and quickly rinse in cold water to remove any salt that maybe on the outside – we don't want our delicious contents to be contaminated! Your apprentice (or apprentices as I am sure once the first batch has been made everyone else will want to make their own!) can eat the ice cream straight from the bag with a spoon or the contents could be squeezed into a cone and decorated with a chocolate flake. Yum! Bon Appetite!

Science in a Nutshell

Pure water freezes, or melts, at 0 °C and boils at 100 °C. So between 0 °C and 100 °C, water exists in the liquid state. It's molecules are provided with enough heat energy and hence kinetic (movement) energy to move around, but not enough energy to break the relatively loose, 'sticky', bonds between them. When you pour liquids into containers you will notice that they all flow and change shape to fit the dimensions of the base of any container you may put them in. The shape of a liquid can change but its volume, at specific temperatures and pressure, always remains the same; you can test this by pouring a liquid into lots of different shaped measuring jugs: tall, thin, wide and short; the volume is always the same.

However, when the temperature is lowered to below 0 °C, the molecules cease to move around – they vibrate only - and they form the crystalline structure of ice, in which the molecules are held together by stronger, 'more sticky', bonds.

When any substance freezes, the particles within it arrange themselves into an orderly pattern. This arrangement is called a crystal. When table or sea salt (sodium chloride NaCl) is added to water, a saline solution is formed and the forming of this solution interferes with the orderly arrangement of the particles in the crystal. The result of this is an increase in heat energy required to be removed from the solution before freezing can occur i.e. the solution freezes at a temperature lower than 0 °C; Salt acts as a freezing point depressant.

The movement energy of the molecules in a substance is related to the temperature. If the molecules initially have a lot of kinetic (movement) energy and we then remove heat from the substance, the molecules will then also lose kinetic energy; the less kinetic energy they have the lower is the temperature. By adding salt to the water more heat energy must be removed before the solution can freeze and, further more, the more particles of salt added, the more kinetic energy must be removed from the solution before it freezes, in other words: the greater the concentration of the salt (solute) the lower will be the freezing point of the water (solvent).

Sea salt is better that table salt because large crystals dissolve more slowly than small crystals. This allows time for the ice cream to freeze more evenly – it will be smoother. For a solution of table salt in water, under controlled laboratory conditions, the freezing point of water has been measured at -21 °C. In the 'real' world say on pathways, roads and steps leading up to front doors; sodium chloride can melt ice down to -15 °C.

Ice has to absorb heat energy in order to melt, in this demonstration; the heat energy is absorbed from the milk, the surrounding air and the hands holding the bag. When you add the salt to the ice, it lowers the freezing point of the water-salt solution. To melt the newly formed ice which is at a temperature now less than 0 °C, even more energy has to be absorbed from the environment in order to make it melt. The newly formed ice is now colder than before, which is how your milk mixture freezes into ice-cream so quickly and it is for this reason that hands need to be protected from frost bite.

Some parts of the world experience very cold winters and temperatures. On a daily basis they have temperatures very much below 0 °C. In these conditions you can't add salt to the roads and pathways because there is no liquid water present to initially dissolve the salt and cause it to dissociate i.e., go from one NaCl molecule into two ions: Na^+ and Cl^-.

Sodium Chloride, NaCl, isn't the only salt used in de-icing, nor is it necessarily the best choice. NaCl dissolves into two types of particles; one Na^+ ion and one Cl^- ion per sodium chloride molecule. A compound that yields more ions into a water solution would lower the freezing point even more. For example calcium chloride, $CaCl_2$, dissolves into three ions one Ca^{2+} ion and two Cl^- ions. The lowest street measured temperature using calcium chloride is -29 °C as compared with the -15 °C for sodium chloride.

Going Further

On freezing water molecules rearrange themselves into hollow rings. This is why water expands (thus becomes less dense) when it freezes unlike other liquids which contract – hence solid water (ice) floats!

Adding salt to water doesn't only affect the freezing point of water; the boiling point is also raised. This means foods cooked in salty water will cook faster because the water boils at temperatures above 100 °C.

Bubbleology

A warning comes with this section – beware Fun is Guaranteed!

Bubbles allow us to not only explore shape and space, colour, light, properties of air and evaporation (to name but a few) but the trials and errors of finding the key ingredients which make up a super-dooper bubble solution also leads to great investigative science opportunities. Bubbles also, most importantly, allow us to work with a very simple product which absolutely fascinates children (of all ages!) and the making of big bubbles so totally gives the all important 'Wow!' factor.

Bubbleology is most definitely a teaching experience in which you set the scene and then step back, watch the faces of your wards, and be rewarded with the knowledge that you are responsible for the joy and amazement on their faces.

Nuts & Bolts

Bucket with lid to store the bubble mixture
Standard bubble wands
Straws
Pipe cleaners
String
Old badminton racquets with strings removed
Metal hangers
Electrical tape
A4 paper
An array of kitchen utensils
Water pistols!
And..... The Bubble Solution

And talking of Joy – nearly all of the websites and books linked to bubbles list the American washing up liquids Joy[1] or Dawn as their key ingredient. Unfortunately we can't buy these brands in our local shops. But don't despair this is where the fun comes in – set this as a problem for your science apprentices to investigate: What is the best local commercial washing liquid for making bubbles?

The following are examples of bubble recipes you might want to try. I prefer the original Fairy Liquid, whilst my good colleague Mark McCluney[2] – Northern Ireland's very own bubbleologist – prefers Cussons. Try both. Let me know what works best for you!

Recipe 1: 1 part washing up liquid, 15 parts distilled or rain water, 0.25 parts glycerine.

Recipe 2: 1 part washing up liquid, 10 parts distilled or rain water, 0.25 parts glycerine.

Recipe 3: 2 parts washing up liquid, 4 parts glycerine, 1 part white karo syrup.

Splooze's recipe: 1 litre Fairy Liquid, ¾ filled bucket of distilled or rain water, ½ cup glycerine.

Secrets for Success

The first step is to have a suitable bubble mixture. If money is no factor you may wish to buy ready made bubble solutions. These are available from shops such as Tescos, Woolworths or the Early Learning Centre in 1 or 2 litre bottles, though only usually in the spring or summer time.

Glycerine can be bought in chemists. Smaller bottles can be bought in supermarkets in the baking section. This is where liquid glucose is also found.

It is very important, if you are making your own bubble mixture to leave it for a couple of days before playing with it.

Make sure your bubble wands and anything your bubbles may touch are wet.

Just slowly dip your bubble makers into the solution – don't slosh them around in the solution – this creates suds and foam, which are bubble busters!Look for cool, humid days and shady areas. Avoid windy days!If you get lots of small bubbles instead of one big one, try blowing more gently or move the wand away from your mouth.If making a bubble by moving the wand through the air, finish your bubble with a quick twist of your wrist to seal it.If you know that your tap water is soft, then tap water is fine, but hard water contains lots of salts, and possibly iron dissolved in it. These are bad for bubbles! So use distilled or rain water instead.

Testing the solution

Make your own bubble blowing tube using the A4 paper[3]. Use this tube to test your solutions. Do this indoors. Dip your blowing tube into the solution. Blow a bubble exhaling ten times. After the tenth breath cover the mouth end of the tube, to prevent air escaping, and start timing the life span of the bubbles; when the bubble bursts, record the  lifespan time of the bubble in minutes and seconds. Do this three times. Sum the three times and divide by three to calculate the average life span of the bubbles. My bubbles using Fairy Liquid lasted 1 min 36 seconds. Using recipe 1 and the revered joy as the washing up liquid my bubbles lasted for 1 min 49 seconds – not that much difference.

Playing

Having waited for two days for the bubble solution to settle your science apprentices will be biting at the bit to start playing. To save disappointment, I advise that you heed the secrets for success, believe me an extra day's wait would be worth it!

Below are some activities that you can do with the bubble solution, but initially just try dipping, blowing or waving your bubble wands in the air, this can on its own bring many hours of fun to the young. Water pistols? Well why not? Have an apprentice make a bubble and shoot it down – obviously if pistol is not aimed at the bubble the possibility of a water fight is relatively high!!

Make wiggly worms.

Blow a bubble and catch it on the bubble wand. Hold it upside down. Dip the end of a short, fat straw into the bubble solution. Blow through the other end. Attach this bubble to the end of the hanging bubble. How many bubbles can you get in a row before they all burst? Wiggle the wand. Watch the worm dance!

The cube

An amazing exception to the rule! (Well not really.) When a normally round bubble is surrounded by other bubbles, it can be made to take on a seemingly odd shape. Surround a bubble by six others. It will appear to be a bubble cube. If the surrounding bubbles are popped, though, the bubble in the center will revert to its natural "round" shape. To make the inner bubble more visible you can blow some smoke into it using a straw dipped in bubble solution.

Look at shapes

Let's play with bubbles. Make bubble wands out of objects found at home or in the classroom. Pull a metal hanger into the shape of a loop, wrap tape around the squashed hook to form a handle or thread string through straws then tie the ends to form a closed loop.

Make cuboids or pyramid shaped frames using straws and pipe cleaners. Use the straws for the straight pieces. Connect two straws by inserting a doubled up pipe cleaner into the end of each straw. In places where three straws meet, fold the pipe cleaners three ways. Attach a pipe cleaner handle to your frame.

There is a myriad of things that you can use to dip into the solution, potato mashers, spoons with holes, the plastic piece used to hold cans together, etc. Look at the fascinating geometrical shapes and colours that the soap films form. Very gently blow and watch the surfaces gently undulate.

Fancy a game of squash?

Remove the strings from an old squash racquet. Dip it into your bucket of soap solution. The strings are replaced with a soap film. To play squash you need a ball; blow a small bubble and bounce it on the soap film! After a few bounces swoop the racquet upwards, around the bubble to trap the 'bubble ball' inside. If the racquet is too ragged wrap some band aid tape around it. The added cloth will also hold more liquid and will allow you to make bigger bubbles.

Science in a Nutshell

Why do we need soap to make bubbles?

It is impossible to blow bubbles with just water because the surface tension in water is just too strong for bubbles to last. By adding soap we decrease the pull of surface tension, typically to a third of that of pure water. Soap molecules are composed of long chains of carbon and hydrogen atoms. At one end of the chain is a group of atoms which like to be in water and at the other end a group which are hydrophobic. However even though they don't like water they love grease which is why soap is used in washing-up liquids and washing powders.

In a soap-and-water solution the hydrophobic ends try to escape from the water and those that successfully manage to squeeze their way to the surface do so by pushing water molecules away from each other. This means the distance between the water molecules increases, hence the surface tension which is caused by the attractive forces between the water molecules, decreases.

The soap film is somewhat protected from evaporation because the hydrophobic ends of the soap molecules stick out of the surface of the bubble.

Why do bubbles pop?

Evaporation, air turbulence and contact with a dry surface are the main reasons for bubbles bursting. Glycerine is used to help slow down the evaporation of water. Liquid Glucose is cheaper and does a similar job though I recommend that it be omitted if you live in an area with lots of flying insects – they will just love it!

If you are playing with bubbles indoors a humidifier is very useful as bubbles don't like dry atmospheres I bought mine using e-bay.

To catch a bubble you must dip your hand or bubble catcher in the soap solution.

Why do bubbles have colour?

One of the most beautiful aspects of bubbles is their colour (which is enhanced with the addition of glycerine to the soap solution). We see the colours through the reflection of light waves from the inner and outer surfaces of the bubble wall. A ray of light that reflects from the inside surface of the bubble travels a slightly longer distance than a ray of light which is reflected from the outside surface.

On route to the bubble these rays are in phase, they are 'in-step' with each other. However after reflection, when these rays recombine they can get 'out of step' with each other and the interference that takes place is responsible for what we know as colour. The amount they are out of step would be equal to double the thickness of the bubble wall – this is obviously a very small distance. How could this distance be measured?

If the waves are traveling in phase and reinforce each other, the colour is more intense. However if they are in anti phase the waves get close to canceling each other out, and there is almost no colour. As the bubble wall gets thinner, the out of phase distance decreases and the two reflected waves of light start to coincide and cancel each other out. The result is that the bubble loses its colour and can become nearly invisible.

If you let a bubble hang from a bubble wand for a while, the interference colours begin forming stripes, because the bubble film is thicker at the bottom (gravity has pulled the solution to the bottom) than at the top, forming a wedge shape. As the bubble drains, the wedge of bubble solution gets thinner and thinner. The black film which then appears at the top is a harbinger of an up-and-coming disaster. The bubble is now so thin it will...POP!

Show your apprentices that you are a 'pop predictor!'

Blow a bubble. Watch the top of the bubble closely. When a black band begins to form on the top of the bubble, announce that it is ready to pop! Or for an even more whacky approach use your light making and noise whooshing wand; at the point of popping aim the wand at the bubble, press the button and...Whoosh! POP! Magic!

Why are bubbles round?

Bubbles and balloons have a lot in common. Scientists refer to them as 'minimal surface structures.' Like a balloon the surface of a bubble is stretched and so it stores up a little bit of energy and like bubbles it also picks a shape which gives it the least amount of stretch. This shape will also have the least amount of surface area and the least amount of stored potential energy. The shape is spherical because the geometric form with the least surface area for any given volume is always a sphere, not a pyramid or a cube or any other form. A sphere. Don't believe me! Test the fact.

1   Joy can be purchased from www.butterfingers.co.uk

2   Mark McCluney http://www.cahootsni.com/e_ed.htm

3   Instructions on how to make your own bubble blowing tube:

4   http://www.zurqui.co.cr/crinfocus/bubble/tube.html

Slimey Encounters

Nuts & Bolts

Cornflour
Food colouring or paint
Large mixing bowl –
preferably transparent
Water
Lots of apprentices willing to get messy!

Secrets for Success

You can either, make one large batch of slime and let your apprentices play with the results, or you can get them to make their own batch of slime – which is much more fun for them! The following is a set of instructions which can obviously be tailored to the ability of your young scientists or to your ability to tell stories!

Pour some cornflour into a mixing bowl. Stir in small amounts of water until the cornflour has become a very thick paste. To make the slime the colour of your choice, thoroughly stir in about 5 drops of food colouring to the mixture.

Stir your slime REALLY slowly. This shouldn't be hard to do.

Stir your slime REALLY fast. This should be almost impossible to do.

Now punch your slime, not too hard though because it will feel like you are punching a solid!

Take a handful of slime and knead the mixture, form it into a ball whilst continuously keeping pressure on it. Open your fingers and remove the squeezing pressure. The mixture will flow like a liquid through your fingers. Yuk!!

Put you hand back into the bowl and very slowly let your fingers sink to the bottom of the bowl. Tickle the bottom – Feel something? Maybe the dreaded slime-living monster? Pull your hand away as quickly as possible. Oh NO! It has you!! The Slime and bowl will both move with your hand. This little piece of drama works really well on little girls!

Take another handful of slime. This time split it into two. Keep the pressure on it at all times. Now try to juggle with it! What does it look like as it flies through the air?

Though it looks messy – when left to dry this slime is very easy to clean up or brush off. Nice furniture?? Then this is an activity for outdoors!

Science in a Nutshell

The cornflour slime mixture is sometimes a solid and sometimes a liquid. That's because cornflour doesn't really dissolve; it only forms tiny solid pieces that hang suspended in the water. This strange type of liquid is called a colloid.

Colloids do weird things. The harder you press them the firmer they feel. But when you remove the pressure and open your hand, they run and drip. The secret of handling colloids is this: slow for flow, hard for solid.

When you put pressure on the colloid, the water between the outer cornflour particles flows to the centre of mass. The outer cornflour particles are now in contact; the lubricating water has been removed and it is the frictional force between them which is responsible for the solid nature. When the pressure is removed the water flows back between the particles – hence the slime behaves like a liquid. Other examples of colloids are: fog, whipped cream, foams, jellies and styling gels.

An everyday example of this would be footprints in damp sand. When you press your foot down, the sand around your foot print is seen to dry out. You can demonstrate this phenomenon by putting some damp sand into a clear plastic drinks bottle. When left on the bench there should be a very thin layer of water on the top. Gently squeeze the sides of the bottle. You will observe the water moving downwards, into the body of the sand. This is a strange sight as you would intuitively expect the water level to rise. The sand at the top now looks dryer. When you release the pressure the water returns to the top of the sand.

On a larger scale you could half fill a clear plastic bottle with pepper corns: these will represent the sand or cornflour (large molecules) and then add some table salt: this represents the water molecules which are much smaller. When the sides of the bottle are pressed you will see the peppercorns move outwards and upwards allowing the salt particles to fall down in between them. The outer peppercorns are left without any salt between them. They are touching. Unfortunately this demonstration doesn't allow you to see the salt move upwards, back between the large particles once the pressure is removed; but it is a good visual aid to help explain why the water moves inwards rather than upwards when pressure is applied to the sides of the bottle or the slime ball.

If you walk slowly or stand motionless on wet sand your feet will sink below the surface. In one of the 2004 Brainiac episodes shown on TV they had the infamous Mr Tickle, from Big Brother, run over a swimming pool filled with cornflour slime. If he had tripped half way across he would have started to sink with worrying consequences. How could he have been saved?

Anything that flows is called a fluid. This means that both gases and liquids are fluids. Fluids like water which flow easily are said to have low viscosity, whereas fluids like cold honey which do not flow so easily are said to have a high viscosity. Slime is a special type of fluid that doesn't follow the usual rules of fluid behaviour. This is why it is described as being a non-Newtonian fluid; it doesn't obey the Laws of Motion as set down by Newton. When a pressure is applied to this slime, its viscosity increases and the slime becomes thicker. At a certain point, this slime actually seems to lose all of its flow and behaves like a solid.

Scientists use the word isotropy to describe the property of a fluid which becomes more viscous and firm when agitated. Tomato ketchup is another form of slime; it however becomes more fluid as you agitate it. This is why we strike the end of the ketchup bottle (assuming it is glass because we can just squeeze the plastic ones!) to get the ketchup flowing. Some say this is why Heinz put the number 57 on the bottle to mark the optimum spot to hit for ketchup flow! Is this the case?

Well we have come to the end but I thought the book would not be complete without an example of a mental trick.. So here goes...

Mind reading

Try this mind reading trick on your wards.

Ask them to follow these instructions (cautioning them to keep their answers in their head):

1. Pick a number between 2 and 5;

2. Multiply this number by 9;

3.You will now have a number with two digits. Add the two digits together

4. Subtract 5 from this number;

5. Correlate the number you now have to a letter in the alphabet, for example 1 = A, 2 = B, 3 = C, and so on;

6. Do you have a letter? Excellent. Now, pick the name of a country that begins with that letter;

7. Use the second letter of the name of the country and think of a category of mammal that begins with that letter;

8. Think of the colour of that mammal.
When they have done all this, tell them that they are thinking about a grey elephant from Denmark and watch the surprise on their faces as they realise you have successfully predicated the outcome.

How does it work?

Easy - steps 1 to 5 will always give you the number 4: 1 + 8 = 9; 2 + 7 = 9; 3 + 6 = 9 and so on. 9 – 5 = 4, so step 6 will give you the letter 'D'. Most people in the Northern hemisphere will think of 'Denmark' in step 7 (because, besides Denmark, there aren't many well known countries starting with 'D') and then think of 'elephant' for an mammal starting with 'e' again there are very few mammals beginning with e and the elephant is the best known. Finally, all elephants are grey.

It's amazing how often this trick works!

About the Author

Sue's dynamic and innovative approach to teaching and learning has taken her from the classroom where she taught physics and chemistry to a Senior Management position in W5-whowhatwherewhenwhy; one of the worlds' top class interactive science centres; to finally the wonderfully wacky position of self employment!

As a physicist-turned-magician-of-sorts Sue combines her drama training with her passion for science to produce a product which captures the imagination and provides real experiential learning opportunites for both the young and slightly-older alike. Check out www.science2life.com to find out more about Sue and Science2Life.

Printed in the United States
86425LV00001BA